A HISTORY
OF THE
26th PUNJABIS
1857–1923

BRITISH AND NATIVE OFFICERS, 18th PUNJAB INFANTRY, DELHI, MAY, 1859.

Subr. Ruttun Singh, 18th P.I. Abdoolah Khan, Native Adjt., 18th P.I. Subr. Ajudhea Singh, 18th P.I
Capt. H. King, 2nd in Com., 18th P.I. (Malikdin Khel Afridi) Capt. J. Williamson, Comdt., 18th P.I.
Mian Singh, N. Comt., 8th Pun., P.B. Lieut. J. Gardiner, 98th Regt. and Adjt., 18th P.I.

A HISTORY

OF THE

26th PUNJABIS

1857–1923

The Naval & Military Press Ltd

Reproduced by kind permission of the Central Library,
Royal Military Academy, Sandhurst

Published by

The Naval & Military Press Ltd

Unit 10, Ridgewood Industrial Park,

Uckfield, East Sussex,

TN22 5QE England

Tel: +44 (0) 1825 749494

Fax: +44 (0) 1825 765701

www.naval-military-press.com

© The Naval & Military Press Ltd 2006

Printed and bound by Antony Rowe Ltd, Eastbourne

PREFACE TO FIRST EDITION

LIEUTENANT P. S. STONEY has written this history of the Regiment as a memento of the Jubilee Year, 1907, and in the hope, which all of us must help him to realize, that it may assist in fostering *esprit de corps* among all ranks. The book will be translated, issued, and read to the men from time to time.

These pages show that though it has never yet been the good fortune of the Regiment to experience the hardships of a hard-fought campaign, as a whole, yet wherever either officers or men have served in the field, whether as a detachment, or attached to other corps, they have always done markedly well. Yet, if no account of stricken fields, of battles, or of sieges, are to be found in this record, nevertheless, we may read here that triumphs may be won in peace as well as in war. I do not allude only to our many Silver Shooting Trophies, nor to our Signalling successes, worthy of mention though they be, but to what Lieutenant Stoney points out in his closing lines, a steady improvement from the beginning without a single relapse, culminating in the splendid state of efficiency attained during the last years of Colonel Dillon's period of command. This is a triumph of which every Sepoy, as well as every British and Native officer in the Regiment may well be proud, for the men have had their share in its attainment, and it is well they should know that their officers appreciate the fact.

It remains for us all in the future to make it our earnest aim, not only to preserve the standard we have reached, but to improve on it to our utmost ability, so that when in a few years the time comes for an addition to be made to this little volume, its author may be able once more proudly to record " a steady improvement, no relapse anywhere."

In conclusion, I beg to tender my thanks to Lieutenant Stoney on behalf of the whole Regiment for the labour of love he has undertaken.

A. A. E. CAMPBELL,

Lieut.-Colonel,

Commanding 26th Punjabis.

Dera Ismail Khan,
December 25th, 1907.

PREFACE TO SECOND EDITION

Iᴛ has somehow fallen to my lot to complete the history of the 26th up to date. The date I am taking is May, 1923. This date marks, not only the conclusion of a tour on the North-West Frontier, but also the end of the period when the Battalion was known as the 26th Punjabis. Since that date it has assumed a new name, the 2nd Battalion 15th Punjab Regiment.

It has not yet had much time or opportunity to make this new name as well known, or as highly honoured, as the old one ; but it has begun well. In the winter of 1923-24 it secured the Jhelum Brigade Sports Challenge Cup, four handsome challenge trophies in the Rawal Pindi District Assault-at-Arms, and the District Hockey Shield. Moreover, in the summer of 1924 it impressed its identity very favourably and emphatically upon a large proportion of the officers of Army Headquarters, Simla, by the extraordinary smart and clean appearance of the detachment which had been detailed to find the guards at Viceregal Lodge and Snowdon.

These later successes are outside the scope of the present revision, and must be left to be chronicled by a future writer ; but they show that the change of number has not caused any deterioration in the Battalion spirit.

I have had little to do in the actual writing of the major part of the additions. The Great War period was mostly written by Major Anderson and Captain Farwell, who both served with the Regiment almost all through. To them I am much indebted. Brigadier-General Campbell made some useful suggestions regarding revisions of the first chapter, for which I am also grateful.

As will be seen by a perusal of this history the efficiency of the 26th Punjabis is now as high as ever, and not only am I proud to record this, as Colonel Campbell hoped in the Preface he wrote to the First Edition, but I am also immensely proud to have been appointed to the command of such a fine body of officers and men.

It will also be seen, by reading more especially Chapter III, that the Regiment as a whole has now had its opportunity of experiencing the hardships of a hard-fought campaign, and that it has thoroughly vindicated the promise, alluded to by Colonel Campbell, which it showed in minor campaigns and during peace.

My object in completing this history to date is not only, as in my first effort, to improve the *esprit de corps* of the Regiment, but also to provide a more or less living memento of the noble officers and men who gave their lives, or suffered wounds or loss of health, for the sake of their King and Country during the Great War, as well as during other campaigns. It is to them in sincere affection and regard that I dedicate this book. I hope it will keep the memory of their sacrifice alive in the minds of their fellow-soldiers, and of others who read it ; and demonstrate to the relatives of those departed our sympathy and admiration.

P. S. STONEY,
Lieutenant-Colonel,
Commanding 2/15th Punjab Regiment.

JHELUM,
   May 25th, 1924.

CONTENTS

LIST OF ILLUSTRATIONS

LIST OF MAPS

A HISTORY

OF THE

26th PUNJABIS

1857 - 1923

CHAPTER I.

FIRST FIFTY YEARS.

June 15th, 1857
The Regiment, then called the 18th Punjab Infantry, was raised at Peshawar in June 1857, as a depot battalion, from which men could be drafted to the various regiments serving in Hindustan.

Men were recruited from Government levies in Peshawar, from the Sikh constables of the Peshawar Police, from Peshawar and Yusufzai Pathans, from friendly Afridis (Malik Din Khel), and from Sikhs and Dogras of the Lahore and Kangra districts. Abdullah Khan (Malik Din Khel, Afridi) of the Peshawar Police, whom Colonel Edwardes, C.B., had appointed Subadar, did a large share of the recruiting, and some men of the 5th and 6th Punjab Infantry were attached as drill instructors. For the first two or three years of the Regiment's existence, however, there was very little opportunity of doing much steady drill, as it was continually moving or being used on service.

The local Peshawar government, under whose orders the Regiment was raised, appointed Captain H. T. Bartlett, 21st Bengal Native Infantry and Cantonment

1

B

Joint Magistrate, Peshawar, as commandant, and entrusted him with the collection of recruits. He was assisted by Lieutenant I. Peyton, His Majesty's 53rd Regiment, as second-in-command, and Lieutenant T. L. Roberts, His Majesty's 87th Regiment as adjutant.

The Regiment was armed with muskets taken from the disarmed regiments of Bengal Native Infantry.

August Very soon after recruiting was started the standing of the Battalion was altered from a recruiting depot to that of a separate regiment, to replace one of the several regiments of Bengal Native Infantry which had been disarmed in Peshawar, and barely two months after the first recruit had been enlisted (June 15th, 1857), the Regiment had been called out to help against the 51st Bengal Native Infantry, who mutinied.

This episode is described in Thorburn's " Punjab in Peace and War " as follows :—

1857 "On August 28th the lines of the 51st Native Infantry were overhauled by some Afridi levies (18th Punjab Infantry), who freely taunted and abused the unarmed Hindustani sepoys. The work over, the sepoys were ordered to move out to a camp on the British side of the cantonment. Thinking they were to be killed, and seeing the piled arms of a newly-raised Sikh regiment within reach, in a mad rush for self-preservation they made for the arms, fought desperately for a few minutes, then broke and fled towards the Khyber. Then followed a wild chase by Sikhs and Afridis, headed by the colonel of the fugitive regiment, who died next day from sunstroke. Before Jamrud was reached, the last of the 700 panic-stricken sepoys had been shot down. The example, Edwardes reported, sufficed. The other disarmed regiments were paralysed with the sudden retribution."

On this occasion 2 native officers and 4 men were killed, and 2 British officers, including the commandant, and several non-commissioned officers and men, wounded. Major-General Sir Sydney Cotton reported favourably on the conduct of the Regiment in this affair.

The composition of the Regiment on **October 1st** October 1st, 1857, was as follows :—

	Coys.	N.Os.	N.C.Os.	Men.
Sikhs	4	7	35	282
Trans-Indus Pathans	3	6	24	205
Dogras from Kangra and Hillmen ...	2	1	7	129
Hindustanis	—	1	1	17
Punjabi Muhammadans	1	1	28	124
Total ...	10	16	95	757

This was below the authorized complement, and in consequence of the unhealthiness of the end of 1857 it was not till March, 1858, that the full establishment was obtained.

On January 3rd, 1858, Lieutenant J. **1858** Williamson, 49th Native Infantry, was **January** appointed commandant, and served on in that capacity for twenty years. To him, therefore, fell the task of moulding the Regiment and bringing it up in the way it should go, and though he did not actually raise it, we may well look to him as the man who nursed it through its infancy, and thus largely contributed to making it what it is.*

This same month a detachment of 150 men was sent to Campbellpore in charge of prisoners, who were to build a new cantonment there. The Afridis who marched with this party deserted in large numbers, thinking they were to share the fate of the Guides,

---

* Many years later two of Colonel Williamson's sons wrote and asked that their father's name might be given to the Regiment, or at least, that his name be recorded in the Army List. Though the existing Commanding Officer forwarded and recommended this application, officially it was not sanctioned, as Captain Bartlett had actually raised the Regiment.

who had not yet returned from Delhi, and had evidently disappeared completely. Eventually, on the Guides returning with a large amount of loot, they were reconciled to being sent to Hindustan.

In April, 1858, most of the Regiment, **April** *i.e.*, 4 British officers, 15 Native officers, and 465 non-commissioned officers and men joined the Yusufzai Field Force under Sir Sydney Cotton, near Nowshera. They took part in the capture of Chinglai and of Mangal Thana on April 26th and 29th, and in the subsequent attack of May 4th upon the headquarters of the colony of Hindustani fanatics at Satana, which lay on the right bank of the Indus, about six miles above Torbela.

The following extract from Paget and Mason's "Record of Expeditions against the North-West Frontier Tribes," describes the part taken by the Regiment in this affair :—

"As the force approached Lower **98th Regt.** Satana, skirmishers from the regiments **Guide Infty.** as per margin, were thrown forward **9th P. Infty.** against the enemy's position. At the **18th P. Infty.** same time the 2nd Sikhs and the 6th Punjab Infantry were detached from Major Becher's column to move up the mountain which forms the rear defence of Satana.

"The 18th Punjab Infantry, under Lieutenant J. Williamson, supported by the 9th Punjab Infantry, under Captain J. B. Thelwell, having without opposition reached the crest of the mountain above Lower Satana, and having moved northward along the same, and also on a pathway on the side of the mountain, in two divisions, first came in contact with the enemy, and drove them from the main position, which they desperately defended, with considerable loss. The 18th Punjab Infantry would then have carried the second position, also, had not the fire of the 6th Punjab Infantry, under Lieutenant T. Quin, been already pouring into it ; that regiment had ascended the northern spur of the range, thus taking the enemy's position in rear, and the 6th Punjab Infantry, following steadily up with the bayonet, drove the enemy out of this position towards the 18th Punjab Infantry, and a hand-to-hand struggle of several minutes ensued, till every Hindustani in the position was either killed or taken prisoner. . . . The fighting of the Hindustanis was strongly marked with fanaticism ; they came boldly and doggedly on, going through all the preliminary attitudes of the Indian prize ring, but in perfect silence, without a shout or a word of any kind. All were dressed in their best for the occasion, mostly in white, but some of the leaders wore velvet cloaks."

Of the troops engaged on May 4th there were 6 killed, and 29 wounded, out of which the share of the 18th Punjab Infantry was 4 killed, and 16 wounded, including Lieutenant Van der Gucht.

The conduct of the Regiment was warmly praised by the General Officer Commanding in a Divisional Order, and subsequently in a despatch to the Adjutant-General, extracts from which are given below :—

" 1. Major-General Cotton has much pleasure in offering to the officers and soldiers under his command his best thanks for the very important services rendered by them yesterday."

" 2. To the conduct of the 6th and 18th Punjab Infantry under the command of Lieutenants Quinner and Williamson, was the Major-General's attention particularly attracted.

" These troops had really a severe struggle with the enemy, who were defeated by them with severe loss."

" 3. The 18th Punjab Infantry under Lieutenant Williamson, having without opposition succeeded in reaching the crest of the mountain above Lower Satana and having moved northward, came in contact with the enemy on the height called Shah Noorkee Lurree, where they were strongly posted in a small village, and in a stockade, which position was very desperately defended.

" Lieutenant Williamson and his gallant corps drove the enemy from this height, having killed 36 on the hill top, and wounded many more. Lieutenant Van der Gucht was wounded in the thigh in a hand-to-hand encounter, and a Subadar and five rank and file were killed, and 15 wounded.

" The 18th Punjab Infantry, for the first time in action with the enemy, thus highly distinguished itself under its gallant commander, Lieutenant Williamson, who reports most favourably on the spirited conduct of the European officers, Lieutenants Van der Gucht, O'Malley, and Green."

Subadar-Major Abdullah Khan, Subadar Sahib Singh and Havildar Shah Gul were awarded the 3rd Class Order of Merit for gallantry on this occasion.

The Indian Medal with a clasp for the North-West Frontier was granted, in 1869, to all survivors of the troops engaged under Sir S. J. Cotton.

In August, 1858, the Regiment moved from **August** Peshawar, where they had returned in May, to Nowshera, but a flood came down within twenty-four hours of their arrival and destroyed most of the station, so they were moved to Attock and Campbellpore.

In March, 1859, the Regiment marched to **1859** Cawnpore, thence it was ordered to Bushie on the Nepal Frontier, but while on its way re-directed to Damuriaganj and Simir on the Raptee. In September it relieved the 1st Sikhs at Gondah, and in November, headquarters and a detachment 500 strong, joined part of the field force under Brigadier Holdich, C.B. This detachment was also employed in the final operations against the rebels carried out from Passrampore under Lieutenant-Colonel J. Cormick, commanding His Majesty's 20th Regiment.

The thanks of the Government were recorded for these operations in G.O.C.C., dated June 16th, 1860, and in a Government letter, while a detachment employed on special service was more particularly thanked in the same way. Extracts from these despatches giving particulars of the work done are given below :—

" The Commander-in-Chief desires to thank brigadiers commanding the Trans-Gogra and Saugor field brigades, and the corps engaged under them, for their excellent and most successful services.

" To the Trans-Gogra Brigade under Brigadier Holdich was confided the task of shutting up the Nepal Passes, while the Gurkha Force under the Maharajah Jung Bahadur captured or dispersed the last remnant of the rebels, who fled into Nepal in January, 1859.

" This duty has been most effectively done, and Brigadier Holdich has had the satisfaction of transferring to the Civil authorities a very considerable number of rebel chiefs who were captured by the Maharajah, the precautions taken by the Brigadier to prevent their escape when they were pursued by the Gurkhas, having been eminently successful."

1860 The Regiment was re-united in January at Gondah, and stayed there the whole year.

In December, on being called upon to volunteer for China, almost the whole Regiment came forward, but they were not sent.

1861 The Regiment moved in relief to Moradabad in January, and sent a wing to Alighar.

In May the designation of the Regiment was altered from 18th Bengal Native Infantry to 30th Punjab Infantry, but was very soon after again altered to 26th Punjab Infantry.

At the same time it was reduced by two companies, No. 4 Pathan Company and No. 9 Sikh Company being broken up.

1863 On March 7th, 1863, colours were presented to the Regiment by Sir Hugh Rose, C.B., Commander-in-Chief. After the presentation the Regiment was inspected by His Excellency, who directed the Commanding Officer to convey to the officers and men his satisfaction with the way in which the Regiment had drilled.

In October the Regiment marched in **October** course of relief to Gorakhpore.

In 1864 the new organization for Native regiments was introduced, and the number of **1864** British officers raised from 3 to 6. Major Williamson was commandant, and the others were a second-in-command, a wing officer, an adjutant, a quartermaster, and a doing-duty officer.

In August, 1865, the Regiment received orders **1865** for service in Bhutan during the ensuing cold season. They were to provide their own transport train of mules and ponies, and Captain Smith was sent to buy some in the Punjab. They started from Gorakhpore on October 25th, and, embarking in flats and steamers at Sahibganj, proceeded up the Brahmaputra to Gowhatti, and thence by road to Diwangari in Bhutan, where they arrived on December 12th.

On February 4th, 1866, three companies **1866** under Major C. M. Longmore advanced as part of Colonel Richardson's force to Saleeka, where the enemy showed some resistance ; but only two men were wounded, and the force subsequently pursued the enemy to the chain bridge over the River Monass, which was taken and held by our troops. A couple of days later headquarters and the remainder of the Regiment followed this wing, and the whole advanced to Monass Bridge. On February 23rd the Regiment was ordered to move to the front across the Monass Bridge. This movement, however, was hardly completed when the British guns which had been captured by the enemy in the previous year were brought into camp. As the recovery of these was the whole object of the campaign, the troops were ordered to return to Diwangari, and the Regiment then proceeded to Alipore where they were encamped.

THREE EARLY COMMANDANTS.

Col. James Williamson,
1858-1877 (nearly 20 years).

Lieut.-Col. M. G. Smith,
7-8-77 to 5-9-81.

Lieut.-Col. G. K. Chester,
6-9-81 to 15-9-86.

During the campaign in Bhutan the health of the Regiment, notwithstanding much exposure and fatigue, had been generally good, but Subadar Saif-Ullah Khan died of malaria on the return journey, and shortly after arriving at Alipore a very severe type of malarial fever became prevalent. There were also several cases of cholera, which necessitated the Regiment being moved temporarily to Fort William.

1867-1878 For the next few years nothing of much importance happened to the Regiment. It moved in relief to the following places :—

Station.	Duration of Stay.
Mehdipore	... December 7th, 1867, to November 21st, 1868.
Ambala	... January 9th, 1869, to October 5th, 1871.
Peshawar	... November 7th, 1871, to October 31st, 1874.
Mian Mir	... November 27th, 1874, to September 27th, 1878.

1875
November
1876 On November 3rd, 1875, the Regiment marched down to the camp of exercise at Delhi, and took part in the field manœuvres and grand march past before Field-Marshal His Royal Highness the Prince of Wales on January 12th, 13th, and 14th, 1876.

1878 On September 27th, 1878, the whole Regiment proceeded by rail to Multan under orders to march to Quetta on field service. Lieutenant-Colonel M. G. Smith was commandant, and there were 5 British officers, 8 Native officers, 27 non-commissioned officers, and 382 men when the Regiment left Mian Mir. The Regiment marched out of Multan on October 1st, and going via Dera Ghazi Khan, Rajanpur, and the Bolan Pass, arrived at Quetta on November 5th. Here it was allotted to Brigadier-General Nuttall's brigade, which formed part of

Major-General Biddulph's force. They left Quetta on November 22nd for the advance on Kandahar, two companies remaining behind at Chaman as **1879** garrison of that place, and the remainder, arriving at Kandahar on January 3rd, were quartered in the city, where they were kept as garrison for a few months. The Regiment marched back in three separate parties via the Bolan Pass, Sukkur, and Multan, and re-united at Sialkot on April 14th, 1879.

For the Afghan Campaign the Regiment received sanction to wear on its colours the words " Afghanistan, 1878-79," and the men were given a medal and donation batta.

On August 25th, 1880, orders were received to **1880** go to the Kurram valley, and the Regiment, which had since moved to Rawalpindi and Jhelum, collected at the latter place. On November 5th fresh orders came for it to be ready to go to Kandahar, and later on, other orders saying it was to hold itself in readiness to proceed on field service against the Wazirs, but no definite orders to march were given until April 12th, 1881. Then it was directed to proceed to Kohat on the " Kabul Scale " in **1881** ordinary relief of the 5th Punjab Infantry.

On the way to Kohat telegraphic instructions were received to push on by double marches. This was most difficult to accomplish, as the carriage was very limited and mostly wheeled, and the weather very unfavourable. The Regiment, however, arrived at Kohat on April 24th, and stayed there a month, when the conclusion of the Wazir campaign permitted it to return to Jhelum.

During the next few years the Regi- **1882-1885** ment moved in relief to the following stations :—

Station.	*Duration of Stay.*
Peshawar	... June 28th, 1882, to December 20th, 1884.
Meerut ...	... February 21st, 1884, to March 4th, 1886.

While at Meerut it was twice ordered (in April, 1885, and in February, 1886) to hold itself in readiness for field service, but each time the orders were countermanded.

1886
March On March 4th, 1886, however, orders came for it to go to Burmah and join the Burmah Field Force. It arrived at Rangoon on March 15th, and on being requested by the Deputy Commissioner of Henzada to assist against dacoits, the greater portion of the Regiment was sent to scour the country under the several officers. A telegram from headquarters at Mandalay, however, urged a speedy arrival there, and Colonel FitzGerald, Commandant, re-embarked the Regiment (less " A " " E," and " F " Companies, left behind under Major Farwell) and proceeded up-country. Another company under Subadar Mansur Khan was left at Katha, and later on acquitted itself very well in some desultory fighting which it had.

Immediately on the Regiment's arrival at Mandalay, " D " Company, under Lieutenant MacCartie, joined a column under Captain Wace, R.A., at a place called Mausi, where the column was acting against a refractory tribe of Kachins. This expedition, being unsuccessful, had to retire, and Subadar Raja Singh, who covered the retirement, was very highly spoken of by Captain Wace for the manner in which he behaved.

April On April 16th, 1886, Captain Dening with one company was sent out to command the post Mausi, where there was some fighting on several occasions, 4 sepoys being killed.

May On May 8th Lieutenant MacCartie was mortally wounded while making a reconnaissance in the Kachin Hills. One sepoy was killed and 1 was wounded in this affair, and the following havildar and 4 men (the survivors), were rewarded with the 3rd Class Order of Merit :—

Havildar Umr Jan.	Sepoy Said Baz.
Sepoy Nur Din.	Sepoy Nur.
Sepoy Mirbaz Shah.	

The Regiment accompanied a column commanded by Colonel FitzGerald against the Kachin tribe which had repulsed Captain Wace, and though they had fighting, both coming and going, there were no casualties.

1886
December
On December 8th Captain Dening, commanding a detachment of about 60 men at Shwekee Gee, was attacked by dacoits, but after six hours' heavy fighting the dacoits were driven off with much loss. Our casualties were 1 sepoy killed, and 1 wounded. Subadar Lakhmir Singh, who headed two sorties, behaved with great coolness and judgment on this occasion.

Major Dening was afterwards awarded the D.S.O. for services in Burmah, and Subadar Lakhmir Singh admitted to the Order of British India, 2nd Class.

1887
March
On March 2nd, 1887, Subadar Mansur Khan, who was in command of the escort to a party constructing a telegraph line in the district near Bhamo, attacked and stormed with only eight men a village held by dacoits, who had kept a party of 100 Burmah Military Police (Gurkhas) under a British District Superintendent of Police at bay for some hours. For this act of gallantry the Subadar was recommended for the Order of Merit.

Two companies of the Regiment with Lieutenants Dennys and Bailey took part in the expedition against the Woontho Sewaha, and were well spoken of.

During the stay of the Regiment in Burmah it suffered heavily from sickness caused by exposure and inferior diet. Out of a loss of 1 British officer and 133 Native ranks, only 1 British officer and 7 men were killed in action. The Dogras lost as much as 33 per cent, the Sikhs 23, and the Pathans only 7 per cent.

For service in Burmah the Regiment was allowed to have the words " Burmah, 1886-87 " on their colours, and the men were awarded the old Frontier medal and clasp inscribed " Burmah, 1886-87."

The Regiment returned to Meerut in March, 1887, arriving there on April 30th.

NATIVE OFFICERS, MEERUT, 1888.

Jemr. Nur Ahmed   Jemr. Lal Singh   Jemr. Kuthuddin   Subr. Goolab Singh   Jemr. Jiwan Singh
Subr Sadar Din   Subr. Alar Singh   Subr. Lakhmir Singh   Subr. Major Abdullah Khan   Subr. Raja Singh   Subr. Mansur Khan
Jemr. Kesar Singh   Subr. Kirpal Singh   Jemr. Sundar   Jemr. Nadir   Jemr. Ganesha Singh

Thirty-six men of the Regiment had volunteered for and been transferred to the Police in Burmah, and were formed into Mounted Infantry. The officer commanding the Police Battalion, Captain H. O'Donnell (later Brigadier-General, commanding Bannu Brigade) wrote specially to the Commanding Officer commending the good service these men had done. Extracts from his letters may be of interest :—

" The detachment I have here under Havildar Atman Singh (late Sepoy) has done splendid service. I have had two severe fights here, one on April 23rd last, when my mounted infantry did wonders. I had 19 mounted infantry under Atman Singh, whom I sent to my right flank to cut off the rebels' retreat, and I attacked their stockaded position on a hill 200 feet high. I drove them out, and Atman Singh attacked them with only eight troopers, rushing in and out of 200 men, the rest of my troopers had lost their ponies in the row. Atman Singh and his men had only carbines, and they showed wonderful pluck in their gallant charge. In this fight Naik Rur Singh showed up, also Sepoy Zamin Shah."

* * *

" The mounted infantry charge of that handful of Sikhs, Punjabi Muhammadans, or mixed, over a wide paddy plain for two miles quite unsupported, was a grand sight, and after that day's fight I never had a doubt of my force's ability to hold its own. We were then over three weeks away from any assistance. In this charge all stripped themselves and only had their loin-cloths on, hair flying, bandoliers slung, and Snider carbines, which they used as pistols. They returned to me unhurt, each man with two or three heads tied to his saddle by the hair, and each carrying two or three dahs from which they requested they might each select one to wear in lieu of a sword. I asked them why they had stripped, and they said

' to hoodwink the Burmans and Kachins,' which they certainly did.

" They charged right through, used their carbines as pistols, then reformed on the other side, loaded, and repeated the above. The enemy never again dared face us in the plains. Those 60 odd men I got from the old 26th Punjab Infantry, and, I think, No. 8 Mountain Battery, together with some 80 Gurkhas, were the only trained men I had for three years, and to them I owe our success then, and the adding of all the country north of Bhamo to the Raj.

" The proudest day I ever experienced was at the end of the fight in Umgomy, when I, after a short pursuit, came back to the battlefield and found my remnants of 75 (8 killed and 15 wounded) just finishing off the last of those who refused to retire. When Elliott and I came up, every man, Sikh, Punjabi Muhammadan, Afridi, Gurkha, all joined as one man to cheer the two boy-sahibs, and to show their delight at our safety. For a time there was absolutely no class distinction, all praised each other, etc. It was a glorious sight."

* * *

" To your old Subadar-Major (Abdullah Khan), who did not let an indifferent soldier come to me, I owe much."

* * *

" On leaving, Trooper Zamin Shah asked me to take him. I knew I had hard fighting before me, and as I was only too glad for volunteers, I took him. . . .

" In this fight I lost 8 killed and 15 wounded· Zamin Shah, I am sorry to say, received a severe bullet wound through the chest, but not before he had killed two rebels with his bayonet and shot a third. . . .

" Zamin Shah . . . . was my personal orderly, by his own orders, and throughout the time he was with me, stuck to me and was my body-guard. On one occasion

BRITISH OFFICERS, 26th PUNJAB INFANTRY, JHELUM, 1894.

Lieut. H. A. Harington   Lieut. L. B. Walton   Surgn. Capt. J. Close   Lieut. C. Rattray   Lieut. O. H. Lawson
Capt. F. H. Hancock   Capt. H. A. Ravenshaw   Lieut.-Col. W. C. Farwell   Capt. C. C. Anderson   Lieut. I. Thompson

he saved both my life and that of the late Major (then Lieutenant) Elliott, assistant commissioner.

" On this occasion, and before it, for his work in the Umgomy fight, I offered him promotion to Jemadar, but he would take no promotion, and if promoted would (he said) play ' Old Harry ' and get reduced. So he remained a sepoy, or rather 1st grade sepoy, on 16/- p.m., wearing a lance stripe, and was my personal body-guard."

1889 The Regiment stayed at Meerut till October 15th, 1889, and then marched to Peshawar, where it arrived on December 19th. While here Subadar-Major Abdullah Khan Sardar Bahadur was transferred to the pension establishment after over forty years' service, thirty-three years of which was in the Regiment and the remainder in the Police.

1890 He had done excellent service throughout his career, and was particularly well spoken of on service in Burmah, and, as a fitting farewell, a parade was held in his honour when he left. Abdullah Khan just lived to see the Jubilee year of the Regiment, though 1905 was the last occasion that he visited it, coming all the way from Kohat to Dera Ismail Khan, and then riding a camel to Paniala to greet his old officers and Regiment, though he was practically blind.

The composition of the Regiment was changed about this time, two companies of Yusufzai Pathans being replaced by two companies of Afridis. The last of these Yusufzais have only lately (1907) been eliminated, as men were allowed to serve on if they wished.

1892 The Regiment left Peshawar on December 23rd, 1892, and, staying for a camp of exercise in Rawalpindi on the way, reached Jhelum in relief on February 7th, 1893.

1894 This year the Regiment distinguished itself very much at musketry, winning the " Honour and Glory " match with the magnificent score of 921 points, an average of 92.1 per man, though the standing position was used at 200. One hundred and forty-six teams competed, including British and Native cavalry and infantry and various scratch station teams, and last but not least, some very redoubtable Volunteer combinations. Jemadar Magar Singh was the hero of the day, making 98 points out of a possible 105. For this feat the Regiment was given a handsome silver challenge shield, and each member of the team received a miniature silver medallion. The above shield had to be given up at the end of the year, but it was never competed for in India again, and was lost sight of altogether. The 26th were the first Native regiment which had ever won the shield, and they were the last team to hold it.

November Lieutenant I. F. R. Thompson was sent with the 20th Punjab Infantry to Waziristan, where that Regiment formed part of the escort to the Delimitation Party. He was present at the night attack on Wano on November 3rd, receiving a gunshot wound in the arm. In connection with this affair, as matters were assuming a more serious aspect, two companies of the Regiment were ordered to Multan, and eventually, on December 26th, the whole Regiment was moved to Dera Ismail Khan, a depot being left at Jhelum.

1894 The Regiment this year provided the guard for the Resident in Kashmir, at Srinagar, and when this guard, which was composed of Dogras, returned to the Regiment, the Resident wrote saying that the conduct of the detachment had been excellent, and there had not been a single case of breach of discipline, reflecting great credit on Jemadar Nihala, who was in charge.

1895 During most of the time at Dera Ismail Khan the Regiment furnished a detachment at Tank of 100 men, but did not see any service.

NATIVE OFFICERS, JHELUM, 1894.

Jemr. Shiraz   Subr. Umar Jan   Jemr. Nihala   Subr. Mehr Singh   Jemr. ——

Jemr. Golukai   Jemr. Jetha Singh   Jemr. Alam Khan   Jemr. Magar Singh   Jemr. Dewa Singh

Subr. Mohd. Afzul   Subr. ——   Subr. Kirpal Singh   Subr. Goolab Singh   Subr. Kesar Singh   Jemr. Ishar Singh

TEAM FOR "HONOUR AND GLORY" MATCH,
INTER-REGIMENTAL, B.P.R.A., MEERUT, 1894.

Indar Singh   Natha Singh   Havr Lehna Singh   Bugle Maj. Juma
Havr. Makhmad   Subr. Magar Singh   Havr.
Jemr. Ishar Singh   Havr. Harif

On their way back to Jhelum in April the Regiment was ordered to form part of the Reserve Brigade of the Chitral Relief Force, and on April 17th proceeded by train to Mardan. Here they remained under canvas till August 23rd, but were only employed as escort to convoys as far as Jalala, and did not cross the frontier. The Reserve Brigade was composed as follows :—

Rifle Brigade, 2/1st Gurkhas, 2/2nd Gurkhas, and the 26th Punjab Infantry, under the command of Major-General Channer, V.C., C.B., but the Rifle Brigade did not leave Rawalpindi.

The Regiment arrived back at Jhelum on August 24th, 1895.

1896

Egypt

On May 11th, 1896, telegraphic instructions were received to mobilize for service in Egypt. Complimentary messages of congratulation were received from His Excellency the Commander-in-Chief, Lieutenant-General Sir William Lockhart, who had served in the Regiment in Bhutan, and also from Major-General Moorsom, commanding Rawal Pindi district.

The latter expressed himself as follows :—

" While regretting the loss from his command of so excellent a regiment as the 26th Punjab Infantry, Major-General Moorsom congratulates Lieutenant-Colonel Dening, officers, non-commissioned officers, and men on their selection for service, hopes that opportunity of adding to their honours may be given them, and is confident that of such the fullest advantage will be taken."

All ranks were immediately recalled from leave, but unfortunately some of the Afridis were unable to rejoin before the Regiment started, so their places had to be filled by transfers from the 20th and 21st Punjab Infantry, who furnished 3 non-commissioned officers and about 100 men.

The Regiment railed to Bombay, arriving there at 11 a.m. on May 21st, embarked at once on the R.I.M.S.

Warren Hastings and sailed for Suakin at 5 p.m. the same day, that is, only ten days after the receipt of orders to mobilize. After a fairly rough voyage (a new experience to many, who learned the meaning of sea-sickness !), the Regiment reached Suakin on May 30th, but was then ordered to Trinkitat to form the garrison of Tokar and its detachments. They disembarked the following day, and an hour later marched off towards Tokar. They halted four hours at Tab, where a detachment was left, and whence another was detached and sent to Abdallarai, ten miles distant. The march was continued through the night, guided by an Arab, for the track was barely distinguishable, and there was thick jungle all round. After a trying twenty-one miles' march the Regiment reached Tokar— where it was destined to spend five long months—in the early hours of the morning. While at Tokar the Muhammadans were much exercised by realizing that Mecca lay to the east and not to the west.

Tokar Tokar is a collection of mud huts surrounded by a wall eight feet high, outside which again was a deep ditch. After one of the severe dust storms (*simoom*) which are prevalent in the country, the wall and surrounding country were found to be nearly level, and a man could walk over the wall of the fort which had been intended as a formidable obstacle. These dust storms were a source of great discomfort to every one, as they took place almost daily, and one even lasted for fifty-six hours. Besides this hardship there was great difficulty about getting water, which was always the colour of red clay, and the barrack accommodation was wholly insufficient. As was not improbable, therefore, the health of the Regiment was somewhat affected, and scurvy broke out, necessitating an issue of extra rations.

Previous to the departure of the Regiment from Egypt the following message was received from the Secretary of State :—" I am directed by the Queen-

Empress to express her satisfaction at the steady soldier-like conduct of her Indian troops while employed on the unattractive, but important duty of holding Suakin, and the surrounding country, during the recent operations in the Soudan. This service has been performed with an alacrity and cheerfulness which has added to the reputation of Her Majesty's Indian Army, and though remote from the scene of operations in the valley of the Nile, it has sensibly contributed to its remarkable success."

Brigadier-General Egerton, C.B., D.S.O., commanding the Suakin Force, among other complimentary remarks in a valedictory order, mentions his high admiration for the discipline and conduct of the troops which he had the honour to command, and thanks all officers for the ready support they afforded him, and all ranks for their uniformly good and soldierly conduct, which it will be his great pleasure to bring to the notice of His Excellency the Commander-in-Chief in India.

The Regiment left Tokar on November 7th, and arrived at Jullundur on December 1st, followed, two days later, by the depot from Jhelum.

For this service two medals were received ; one, the Queen's Medal, " Soudan, 1896, Dongola Expedition," and the other, a medal presented by the Khedive of Egypt.

**1897
N.W.
Frontier** On August 13th orders were received to proceed to Peshawar on relief scale without families, but taking mobilization equipment, in connection with the disturbances on the North-West Frontier. The Regiment left at 9 a.m. on the 14th, and arrived at Peshawar on the morning of the 16th, being joined the same day by detachments which had been at Ludhiana and Amritsar. All leave and furlough men had rejoined within ten days of receiving information.

One wing under Major Ravenshaw was sent to Jamrud, but very soon moved to Hari Singh-ka-Burj.

Headquarters and the other wing moved out somewhat later to Adazai Bridge, near Shabkadar, but did not take part in the fighting at the latter place. When the Mohmand Field Force, under the command of Major-General E. R. Elles, C.B., moved into the Mohmand country, the headquarters and wing 26th Punjab Infantry were ordered into Shabkadar to act as troops at the base. Colonel L. Dening, D.S.O., was appointed base commandant, and Captain L. B. Walton, base staff officer. " G " Company (Afridis), under Captain F. H. Hancock, was attached to the 20th Punjab Infantry, who were part of the Mohmand Field Force from September 15th, 1897, to October 1st, 1897. The wing remained at Shabkadar busily employed in escorting convoys, garrisoning posts, and making roads, from the middle of September till October 7th, when the return of the Mohmand Field Force liberated it.

October By the middle of October both wings were back again in Peshawar, and three companies of Afridis and the recruit training staff were despatched to Mian Mir for garrison duty there under Captain L. B. Walton.

Lieutenants Harington and FitzGerald had been attached to the 38th Dogras during the operation in the Malakand, and, unfortunately, the former was mortally wounded in the Markhanai Valley affair against the Mahmunds on September 16th and 17th, and died twelve days later.

For operations on the North-West Frontier, 1897-98, 8 British officers and 813 Native officers, non-commissioned officers and men became entitled to a medal and gratuity.

Lieutenant-Colonel L. Dening, D.S.O., was mentioned in despatches for services rendered at the base.

1898 In March headquarters and the remainder of the left wing moved from Peshawar to
Jullundur Jullundur to join the Afridi companies which had moved there, but the right wing

under Major Ravenshaw remained on in Peshawar till December, 1898, when it rejoined headquarters.

1899 Major G. F. H. Dillon was appointed commandant in place of Colonel Dening, appointed colonel on the staff.

1900 **Malakand** On January 19th the Regiment proceeded by rail from Jullundur to Nowshera, and thence by route march to Camp Khar to join the Malakand Field Force, commanded by Brigadier-General A. J. H. Reid, C.B.

The double company system was introduced into the Indian Army this year, and, in order to make each double company of one class only, one company of Dogras was transferred to the newly raised 41st Dogras, and the establishment of Afridi companies reduced from three to two. To take the place of these two companies the Regiment was ordered to enlist Punjabi Muhammadans.

The Regiment this year won a hockey tournament at the Malakand, beating the 15th Sikhs, who had won the newly-instituted Punjab Native Infantry Hockey Tournament at Lahore a month previously.

1901 **Peshawar** On November 14th the Regiment left the Malakand and marched to Peshawar, where it arrived on November 20th. On this march, just as the Regiment was coming into Mardan, the old Grunthi, Sahib Singh, who had been in the Regiment over thirty-five years, and was much reverenced by everyone in the Regiment, and by all Sikhs in the Native Army, fell dead at the head of the Regiment where he was marching in accordance with his usual custom. His death was a sad loss to the Regiment.* Few will now remember the grand old figure marching stoutly along in spite of his age, but his memory will remain as the embodiment of loyalty and zeal. A man of simple unswerving faith, he was an example to us all.

In the finals of the Punjab Native Infantry Hockey Tournament we were beaten by the 15th Sikhs by one goal.

* His son now reigns in his stead (1923).

22

1902 The Regiment took part in the Yusufzai manœuvres of 1902, which took place round about Mardan in February.

1903 The Regimental signallers, who had been specially selected, on account of their high reputation, to be employed with the Seistan Boundary Commission in January, 1903, returned again to India in June the same year. Their work had been particularly arduous, owing to the long marches and shortness of water on the arid sandy plains of the Helmund Valley and N.W. Baluchistan ; and, though this service was not reckoned as field operations, it approximated closely to conditions of field service. On the completion of the task they were specially commended.

The Regiment attended the Rawalpindi Punjab manœuvres in December, 1903, and the men worked very well and willingly, in spite of the fact that the marches were long and tedious, and the hardships very considerable.

Lieutenant-Colonel Dillon was made a C.B., and Subadar-Major Magar Singh admitted to the 2nd Class Order of British India, with the title of " Bahadur."

Captain Lawson and Subadar Muhammad Akbar accompanied the Tibet Mission, the former as transport officer, and the latter in command of a section of mounted infantry.

The title of the Regiment was altered from " 26th Punjab Infantry " to " 26th Punjabis," and Major-General L. Dening, D.S.O., was appointed honorary colonel of the Regiment.

1904

Miran Shah On January 14th the Regiment left Peshawar for Miran Shah in the Tochi Valley by route march, going via Kohat and Thal and thence across the hills to Idak in the Tochi Valley. This was not a recognized route, and special sanction had to be obtained to use it, as part of it was over unprotected area. (See Map No. 4.)

BRITISH AND NATIVE OFFICERS, PESHAWAR, DECEMBER, 1901.

Back Row : Subr. Ishar Singh, Lieut. P. S. Stoney, Subr. Shiraz, Lieut. G. O. Turnbull, Capt. I. F. R. Thompson
Lieut. A. J. H. Grey, Subr. Allah Dad
Middle Row ; Subr. Nihal Singh, Capt. G. Walton, Subr. Umarjan, Lt.-Col. G. F. Dillon, Subr. Maj. Magar Singh
Maj. F. H. Hancock Jemr. Makhmad
Front Row : Jemr. Yar Mhd Khan, Jemr. Kala Singh, Jemr. Tola Ram

BRITISH AND INDIAN OFFICERS, 26th PUNJABIS, DERA ISMAIL KHAN, JANUARY, 1906.

Back Row : Jemr. Ali Haidar, Subr. Yar Mhd. Khan, Jemr. Tura Baz, Subr. Ishar Singh, Lieut. W. Tarr, I.M.S., Jemr. Mal Singh Subr. Harnam Singh, Jemr. Basimullah

2nd Row, Standing : Lieut. H. T. C. Ivens, Lieut. R. T. G. Salusbury, Capt. O. H. Lawson, Lieut. G. O. Turnbull, Adjt., Lieut. O. D. Bennett, Lieut. I. M. Little, Subr. Mhd. Akbar, Subr. Sham Singh

3rd Row, Sitting : Maj. L. B. Walton, Subr. Makhmad, Subr. Maj. Magar Singh (Sardar Bahadur), Lieut.-Col. G. F. Dillon, C.B. Comdt., Subr. Lahrasap Khan, Subr. Lachman Singh, Maj. L. C. Dunsterville, 2nd in Command

4th Row, on ground : Lieut. R. J. Cargill, Lieut. E. A. Maude, Jemr. Mhd. Quresh, 2/Lieut. G. W. Anderson, Lieut. P. S. Stoney

The Regiment left Miran Shah on October 15th, 1904, after eight and a half months' stay. It was relieved by the Northern Waziristan Militia, who took over all the posts in the valley, all regular troops being withdrawn. The Regiment marched through Bannu to Dera Ismail Khan, where it arrived on October 25th, 1904, relieving the 45th Sikhs.

1905

D. I. Khan

The Regiment took part in brigade manœuvres near Paniala and Pezu in February, being inspected in accordance with Lord Kitchener's test. The conditions under which they were tested were abnormally severe, the weather being very wet and cold, and yet the Regiment did very well; but they did not secure first place in the brigade, being just beaten by some 12 points by Coke's Rifles, who were later adjudged to be the best Regiment in the Punjab.

In March Colonel Harman, commanding the Southern Waziristan Militia, was murdered by a man of the Militia at Wano, and immediately on receipt of the news in Dera Ismail Khan orders were issued to mobilize a flying column to proceed to Wano if required. The Regiment received its orders at 7 a.m. to be ready to move by 10 a.m., but no transport was available. Two days later transport of sorts having been procured a column under Colonel Dillon, C.B., in which the Regiment was included, marched to Tank, completing the forty-two miles in twenty-eight hours. The column, however, got no further than Tank, and after sitting there for three weeks, during which time it rained almost incessantly, returned to Dera Ismail Khan. The whole matter was settled by a Jirgah at Jandola, at which Colonel Deane, Chief Commissioner, North-West Frontier Province, met the erst-while outlaw, the Mullah Powindah, who acted as spokesman and representative for the Mahsuds.

While in Dera Ismail Khan the Regiment took its turns with others in supplying detachments at Jandola,

Zam, Jatta, and Drazinda (in all nearly 300 men), for periods of two or three months. At Jandola there were 2 British officers, but Native officers were in command of the other posts.

1906 The Regiment took part with the brigade in a tour through Sheranni country in January, after the usual inspections carried out from a concentration camp at Rori. The Sherannis had been showing signs of unrest, and the political authorities thought the sight of some troops in the heart of their country might have a salutary effect. The march was carried out under service conditions, but no contretemps took place, and the inhabitants all through appeared quite friendly. The route taken went close under the Takht-i-Suleman and came out by Drazinda and Draband.

Colonel **Dillon** On August 4th Lieutenant-Colonel G. F. H. Dillon, C.B., was appointed staff officer for Militia to the Chief Commissioner, North-West Frontier Province, but had hardly taken over the duties before he fell ill. He proceeded on sick leave to Europe in September, but to the very great regret of the Regiment, and the loss of the service, died at sea between Port Said and Marseilles on September 29th. Colonel Dillon had nearly completed seven years in command of the Regiment, and had very much endeared himself to all ranks by his sterling qualities.

Major A. A. E. Campbell, 25th Punjabis, was appointed commandant in place of Colonel Dillon, and took over command on October 29th.

Sub.-Major **Magar Singh** In November the Regiment suffered another loss in Subadar-Major Magar Singh Bahadur, but in this case it was only retirement to enjoy a well-earned pension, after nearly thirty-four years' service in the Regiment. On the 10th he was given the honour of a special parade, and the Regiment marched past him and advanced in review order giving him the

REGIMENTAL SIGNALLERS, 1906.

Natha Singh   Ishar Singh   Suraj Din
Naik Pal Singh   Mir Dast   Sharm Singh   Maggar Singh   Sher Singh   Dewa Singh   Buta Singh
Madaman   Havr. Kesar Singh   Capt. G. O. Turnbull   Naik Buta Singh   Havr. Attar Singh   Narain Singh
Mul Singh   Bagh Ali

salute, and on the 12th escorted him in a body to the outskirts of Dera Ismail Khan to see him off.

1907

Jubilee

In 1907 fell the 50th anniversary of the raising of the Regiment. To commemorate the occasion a fund was set on foot to assist sepoys going on two months' leave to their homes, by defraying their railway fares.*

This year this could be only partially done, as the fund at the outset was small, but there is reason to expect that soon every man going on long leave will be able to do so free, just as he already does when he proceeds on furlough. The benefit thus accorded to the men of the Regiment should indirectly aid recruiting for the Regiment, and, consequently, conduce to its efficiency.

Major-General R. B. Adams, V.C., C.B., and his staff very kindly presented the officers with a beautiful portrait of Subadar-Major Magar Singh painted in oils as a Jubilee present, and Colonel Campbell presented a similar one of Subadar-Major Makhmad.

The Regiment took part in brigade manœuvres at Pezu, where it also underwent its annual inspection.

In November of the year Lieutenant-General Sir Alfred Gaselee inspected the Regiment, and expressed himself as very pleased, especially with the stamp, the bearing, and the practical system of training of the recruits.

Thus the records of the inspections of the 26th Punjabis show a steady improvement from the beginning, with no relapses anywhere, and it may fairly be said that in this Jubilee year of 1907 the Regiment has reached a state of efficiency, which it has built up by fifty years of good honest work on the part of both officers and men, and through which it can well claim to be second to none in the Indian Army.

---

* Owing to change in conditions this fund has been altered, and some of the money devoted to a platoon efficiency trophy.

CHAPTER II.

1908-1914.

1908
Kohat
The beginning of 1908 saw the 26th Punjabis leave Dera Ismail Khan and move in relief to Kohat. It marched by road through Bannu under the command of Lieutenant-Colonel A. A. E. Campbell, starting on February 17th and arriving on the 25th. Soon after arrival at Kohat, on account of the position of affairs on the Afghan Frontier, orders were received by all units in the Northern Army for the recall of men on leave and furlough. The situation then improved, and within ten days (about May 10th), leave and furlough was again opened. Almost immediately, however, the Mohmand Expedition was decided upon by the Government of India, and the Battalion, though taking no part in it, had to detail 3 British officers (Captain Stoney, Lieutenants Ivens and Salusbury) and 3 Indian officers (Subadar Ishar Singh, Jemadars Kesar Singh and Turra Bay) to join the 53rd and 54th Sikhs. These officers were all commended by their respective commanding officers for good work during the operations. Lieutentant Ivens was dangerously wounded at the top of the hill, above the village of Khan Beg Khor, leading a company of the 54th Sikhs (F.F) in the picquetting operations.

At Kohat the Battalion had an opportunity of recovering somewhat after the unhealthy climate of Dera Ismail Khan, and was able to put in much useful training. In October, 1908, it was re-armed with the M.L.E. Mark I* charge-loading rifle, and was selected to carry out a special course of musketry.

During this year at Kohat the Battalion took its share of the usual alarms and excursions concomitant with life in a frontier station. Raiders from across the border were constantly making inroads into British

26

territory, and robbing or carrying off rich Hindu banniahs. Provided information of such raids was received without undue delay, and there appeared the smallest chance of intercepting the raiders, troops were often turned out to assist the police in pursuit. On one such occasion, December 21st, 1908, a company of the 26th marched out of Kohat at about one and a half hours' notice at 6 p.m. with three days' rations, and marching through the night—a very cold one—reached its destination, about thirty-eight miles away, next morning before midday. Here, with the assistance of a troop of cavalry under Jemadar Hissam-ud-Din, 23rd Cavalry (P.F.F.), and some levies belonging to the Khan of Teri, it immediately laid out a line of outposts stretching across some four miles of country, and watching all approaches night and day for three days, endeavoured to intercept the reported gang of fifteen raiders. At the end of this time, having seen or heard nothing further regarding the gang, the detachment received orders to return to Kohat, where it arrived on Christmas Day, disappointed at its misfortune in missing the raiders, but glad to enjoy a portion at least of the Christmas holidays.

1909 New Colours were obtained early in 1909, and introduced into the Battalion at a ceremonial parade held at Kohat on March 2nd. The old Colours, which were worn out and un-serviceable, were deposited for safe keeping in the officers' mess.

During 1909 the efficiency of the Battalion was more than maintained, as the detail of inspection reports in Appendix II will show. More especially in signalling did it receive great commendation. During the previous two years the signalling reports had not been up to the Battalion's usual standard, but this year the inspector said, " Havildar Buta Singh, the senior assistant instructor, deserves special credit for his individual exertions in the training of this splendid body of signallers, who can have but few equals in the Indian Army."

In consequence of disturbances between the Kurram and the Tochi, a detachment consisting of 2 British officers, 2 Native officers, and 150 rank and file were sent to Thal on March 24th as a reinforcement of the regular garrison. Here it had to stay in tents during a portion of the hottest part of the year, and was not finally relieved till June 26th. Captain Turnbull was in command of this detachment.

Captain Turnbull Captain G. O. Turnbull had just given up the adjutancy of the Battalion, after a tenure of four years. He had devoted himself whole-heartedly and with marked success to the training of the recruits of the Battalion, and had been rewarded by many complimentary remarks made by inspecting officers, with particular reference to recruits : *e.g.*, in 1908, " The training of recruits is specially good." In 1910–1911 Brigadier-General Birdwood, who knew the Battalion well, said : " The men are exceptionally well set up, due a great deal to the great trouble taken with the recruits." That same year Sir James Willcocks said : " The recruits are far away the best I saw in the Northern Army this year." Though at this time Captain Turnbull was no longer adjutant, he had left behind him a standard and a system which assisted Captain Salusbury and subsequent adjutants in maintaining the Battalion's excellent reputation.

Captain Turnbull identified himself also very keenly with the games and athletics of the Battalion, with the result that it generally swept the board at sports meetings in the garrison. In the matter of the men's appearance the period of his tenure of the adjutancy marked a very strict adherence to uniformity and cleanliness, and the Battalion became the envy of other units, especially when turning out in white Hindustani clothing. The custom of uniform white clothing then set by the 26th Punjabis was soon

REGIMENTAL HOCKEY TEAM, KOHAT, 1910,
WINNERS OF PESHAWAR TOURNAMENT.

Bachan Singh   Bhagwan Singh   Yar Akhmad
Kesar Singh   Jhanda Singh   Bishan Singh   Said Ahmad
        Shamir Singh   Capt. G. O. Turnbull   Mehr Khan
        Dial Singh   Hazara Singh

followed by other battalions, and later on was adopted by Army Headquarters throughout the Indian Army. To the Jemadar-Adjutant, Jemadar Jan Gul, one of the smartest instructors in the Indian Army, no small measure of credit would no doubt be attributed by Captain Turnbull.

1910

Inspections

During these years at Kohat the 26th Punjabis earned extraordinary good reports from the various inspecting officers who saw it. His Excellency Lord Kitchener visited Kohat in April, 1908, and expressed himself as thoroughly satisfied that the Battalion was keeping up its old reputation. The following year Sir J. H. Wodehouse, K.C.B., said the Battalion was " particularly fine," and " among the very best he had seen." In 1909-10 and again in 1910-11 Brigadier-General W. R. Birdwood, then commanding the brigade at Kohat, said : " I cannot speak too highly of the general efficiency of the 26th, of which, perhaps, as good a proof as any is the extraordinary respect which all other corps in the garrison have for it."

Sir James Willcocks, commanding the Northern Army, who saw the Battalion, referred to it as : " A really good battalion. . . . . They are a cheery well set up, well disciplined corps. . . . "

Games and

Athletics

The years at Kohat were also marked by the Regiment's excellence at games and athletics. The hockey team, though it did not win the Punjab Native Army Tournament, was one of the best teams competing, and generally won all local matches. There were many stalwart players in those days, many of whom some years later, owing partly to the value of the game as a training, rose to be Indian officers, such as Kesar Singh, Jhind Singh, Mehr Khan, Gonda Singh, Shamir Singh, and Bachan Singh. The officers who contributed to the success of the team were Captain Turnbull, Lieutenants Cargill and Bennett.

At athletics the Regiment did remarkably well,

whenever brigade sports were held. At " khud " racing no unit could find anyone to approach our Afridis, who also succeeded in beating a team of Gurkhas specially selected to oppose them. One, Godar Khan, a Punjabi Muhammadan, was a marvellous runner. The Regimental tug-of-war team, though not particularly heavy, was very successful on account of its good training.

At polo, also, the Regiment produced many good players, who formed a large proportion of the players in station games. From these players a team of four, consisting of Captains Lawson and Stoney, and Lieutenants Salusbury and Bennett, entered for the Native Infantry Polo Tournament, but were beaten by the Guides.

1911
Samana
In February, 1911, after three years at Kohat, the Battalion moved in relief to Hangu and the Samana, being relieved at Kohat by the 56th Rifles (Frontier Force) and replacing the 58th Vaughan's Rifles (Frontier Force). Headquarters and four companies were at Hangu, two companies at Fort Lockhart on the Samana, and two companies in Thal. During the hot weather headquarters and two companies moved up to Fort Lockhart. At Thal, in April, there was a most unfortunate accident, part of the fort wall collapsing and falling on some men sitting below it. Four men were killed, including No. 4829, Pay-Havildar Allah Din, who was a promising non-commissioned officer, marked for early promotion, and who that year had won the Magdala Gold Medal in the annual musketry matches.

1912
Hong-Kong
At the beginning of 1912 the 26th Punjabis had for some months been under orders to proceed to Karachi in relief, and arrangements had already been made accordingly, so it was somewhat disconcerting when, on January 15th, orders were suddenly received to embark at once for Hong-Kong. The recent revolution

in China and consequent internal disturbances were apparently the cause of the unexpected increase to the garrison of Hong-Kong.

The Battalion, under the command of Major Walton (Colonel Campbell being on leave in England), left Hangu by road on January 18th, Kohat by rail on the 20th and 22nd, and embarked at Karachi on board the Indian Marine Ship *Dufferin* on January 27th, leaving behind a depot at Karachi under Captain I. M. Little, who had rejoined from the Cantonment Magistrate's Department.

Hong-Kong was reached on February 13th, and there the Battalion first went into camp on the sea-shore at Kowloon, alongside the 25th Punjabis and 24th Hazara Mountain Battery, who had also been suddenly despatched from India. This site was not comfortable, and the Battalion was glad to move to Rennie's Mill on the mainland on May 22nd. This was a disused flour mill, and afforded a capacious barracks for the men, the officers being in cubicles. The site was rather restricted by the fact that there was a high hill at the back of the mill, and the sea close in front of it. The sea afforded opportunities for bathing and boating, and the majority of the men learned to swim, while many were also instructed in rowing in two eight-oared whalers which were purchased.

Colonel Campbell Colonel A. A. E. Campbell completed his tenure of command on May 9th, 1912, and handed over to Major L. B. Walton, who was appointed commandant in his place. Colonel Campbell came to us from the 25th Punjabis, and at once set himself to improve the efficiency of the Battalion as keenly as if it had been his own. He had been an expert in musketry, having served on the staff as an instructor, and utilized his special knowledge to good effect in the 26th Punjabis. Though the Battalion had for a long time been one of the best shooting Battalions in the Indian Army, it more than maintained its reputation, and this in spite of the fact that during this period musketry competitions developed into much more practical concerns than the former specialized bull's-eye shooting carried out at Meerut.

The Battalion's achievements during these years included the following :—

May, 1907.—4th Double Company (Afridis), Cawnpore Woollen Mills Cup, 1st.

April, 1907.—G. Coy. (Afridis), Commander-in-Chief's Cup, tied for 1st place, but lost on re-shooting.

1908.—G. Coy. (Afridis), Commander-in-Chief's Cup, second.

1909.—Selected men, Commander-in-Chief's Cup, second.

1909.—G. Coy. (Afridis), Cawnpore Woollen Mills Cup, second.

1909.—Chance team, Empire Cup, 10th.

1910.—Havildar Allah Din, Magdala Gold Medal, 1st.

1912.—Chance team, Empire Cup, 14th.

Colonel Campbell was later on appointed to the command of the 10th Brigade at Quetta, and after completing his tenure of that appointment retired as Brigadier-General in 1919.

1913 Training was difficult to carry on at Rennie's Mill, owing to the cramped terrain available. On January 1st, 1913, the Battalion proceeded to Camp Sun Wai, where double company training was carried out from January 1st to February 8th, and regimental training from February 11th to 21st. From March 3rd to the 6th the Battalion took part in brigade manœuvres with the 24th Mountain Battery, 25th Punjabis, 126th Baluchistan Infantry, and 8th Rajputs.

1914 The following year training was again carried out in camp, but this time at a place called Hoshung Heung. In March, 1914, combined training with the Royal Navy and Port Defences was carried out at Hong-Kong, the Battalion taking part in combined Naval and Military

BRITISH AND INDIAN OFFICERS, KOWLOON, CHINA, 1914,

Back Row: Jemr. Sher Akhmad, Jemr. Pala Singh, Capt. Hodge, I.M.S., Jemr. Firoz, Jemr. Suraj Din, Jemr. Tura Baz
2nd Row: Jemr. Shamir Singh, Jemr. Harnam Singh, Lieut. J. D. Fulton, Subr. Sham Singh, Lieut. J. E. Shearer, Subr. Jan Gul, Capt. H. T. C. Ivens
3rd Row, seated: Subr. Mal Singh, Maj. I. F. R. Thompson, Subr. Yar Muhammad Khan, Lieut.-Col. L. B. Walton, Comdt., Subr. Maj. Muhammed Akbar, Maj. O. H. Lawson, Subr. Harnam Singh
4th Row: Jemr. Kesar Singh, Subr. Ishar Singh, Capt. I. M. Little

manœuvres covering several days. These manœuvres were very strenuous, and comprised an attempted landing on Hong-Kong island.

Musketry training was maintained at a high standard, and when firing for the Hatton Cup among the units at Hong-Kong the 26th Punjabis, both in 1912 and in 1913, easily beat the other units, the first time by a margin of over 50%, and the second time by over 30 %.

The Battalion had originally been sent to China as a temporary measure, and expected to be there not more than one year. Therefore at the end of the second year, when orders were received to return to India, all ranks were distinctly glad. Though in some ways they had found China a pleasant change, they had grown tired of the restricted space and the restrictions as regards leave and furlough. Their disappointment was all the more intense when, on April 15th, 1914, these orders were countermanded, and orders were received that the movement to India was held in abeyance. Thus it was that the Battalion spent the summer of 1914 in China.

CHAPTER III.

THE GREAT WAR, 1914-1918.

 On the outbreak of war on August 4th, 1914, the 26th Punjabis found themselves still in Rennie's Mill at Kowloon, near Hong-Kong. They had then been there for two and a half years. No leave had been open for Indian ranks, and even discharges at the expiration of three years (normal period of enrolment) had been stopped.

The first intimation of the possible outbreak of hostilities with Germany was received on July 30th, when the Battalion received orders to adopt precautionary measures for mobilization. On August 2nd 100 rifles of No. 2 Double Company under Lieutenant Dillon proceeded to Pakshiwan. On August 5th news was received of the outbreak of war, and " G " Company under Captain Ivens proceeded to Western Defence Section, Mount Davis.

On August 8th Rennie's Mill was evacuated, the heavy baggage being stored at Holt's Wharf, Kowloon, and the Battalion proceeded on field service scale (20lb.) to occupy some of the Hong-Kong defences. Major Thompson was at this time commanding, as Lieutenant-Colonel Walton did not return from leave in England till November 23rd. Headquarters, No. 1 Double Company, and " H " Company went to " war stations " in the Western Defence Section, " G " Company rejoining the Battalion. Camp was pitched on the University recreation ground. No. 3 Double Company and the remainder of No. 2 Double Company under Major Lawson and Captain Little went to Devil's Peak, Eastern Defence Section.

On August 26th Nos. 2 and 4 Double Companies moved to Victoria Section to replace the Duke of Cornwall's Light Infantry, who had received orders to proceed on service. The remainder of the Battalion also moved there, and with Headquarters at " Sanatorium," found the following posts :—

" High West Gap," " Victoria Gap," " Sanatorium," " Magazine," " Wanchai Gap," " Middle Gap," " Wongmichong Gap," " Taytan Conduit," " Deep Water Bay," " Aberdeen," " Tookfullum," " Tai Ho Wan Cable Hut," and " North Point Cable Hut." On October 12th, Headquarters moved to " Mount Austin " Barracks.

On December 18th No. 2 Double Company rejoined Headquarters, and, two days later, all posts were withdrawn, and all double companies except No. 4 concentrated at Mount Austin. No. 4 Double Company remained at " Sanatorium."

At this time the following officers of the Battalion who were at home in England, were attached to Lord Kitchener's new army :—

Captain G. O. Turnbull, 6th Royal Scots, August 22nd, 1914.

Captain E. A. Maude, 10th Warwickshires, August 7th, 1914.

Lieutenant H. D. Drysdale, 11th Royal Scots, August 22nd, 1914.

Lieutenant L. J. Torrie, 12th Royal Scots, August 21st, 1914.

Captain Anderson and Lieutenant Dillon left Hong-Kong on the s.s. *Delta* on December 3rd, 1914, for field service in Egypt. Of the above officers, Captain Turnbull and Lieutenant Torrie were both severely wounded early in the war, and though they continued to carry on duties during the war, they never completely recovered their physical fitness. Captain Turnbull earned the D.S.O. The two others, Lieutenants Drysdale and Dillon, were both killed in action, the former in France in 1915, the latter with the 24th Punjabis at Ctesiphon, gallantly leading his men on the Turkish trenches, shot down in the act of cutting enemy wire. Captains Maude and Anderson rejoined their own Battalion later on in Mesopotamia.

The desire of the Battalion to be permitted
1915 to prove its efficiency on service and take
its part in the defence of the Empire is reflected
in the inspection report by Major-General F. H. Kelly
commanding South China. He said under " General
Efficiency " " All that can be desired. Regiment
quite fit and longing for service in the field. I hope
after a month's leave in India they will be sent to
France. There is no better fighting unit in the Indian
Army."

It was, therefore, with feelings of great delight that
orders were at last received, early in 1915, for the
Battalion to return to India. Headquarters with
Nos. 2 and 4 Double Companies left Hong-Kong for
India in s.s. *Ellenga* on February 25th, and were
followed on March 12th by Nos. 1 and 3 on s.s. *Assaye*.
The wings arrived at Bombay on March 10th and 25th
and at Agra on March 12th and 28th respectively.

On arrival at Agra orders were received
Agra from Army Headquarters granting one
month's China leave to all ranks in the
Regiment. On the Commanding Officer's recom-
mendation this was extended to six weeks for Afridis
and five weeks for the other classes in the Regiment,
to enable each man to have a clear one month at his
home. All were ordered to rejoin on expiration of
this leave at Agra, when the Battalion would move
to Bannu. Subsequently, when all had proceeded
on leave, orders were received for Regimental Headquarters and Depot to proceed to Bannu, and for men
on leave to rejoin there. The Headquarters and
Depot arrived at Bannu on April 22nd, 1915, and came
under the command of Brigadier-General V. B.
Fane, C.B.

On May 10th, the date on which all men
Tochi were due back from leave, it was found
that though all the Sikhs and Punjabi
Muhammadans had rejoined all right there were 160
Afridis absent. As the Indian officers reported that

these men had no intention of rejoining, the Commanding Officer, on May 12th, despatched a party consisting of Subadar-Major Muhammad Akbar and Subadar Jan Gul to try to induce the delinquents to return. After a couple of months' strenuous endeavours the two Indian officers returned to report their inability to persuade the defaulting men to return. These men frankly refused to serve any longer. They had enlisted for three years in the first instance, and many had completed this period and had been applying for their discharge for several years, while the Regiment was in China. By the time war broke out they were already discontented at being obliged to serve on over the usual three years, and though they had taken an oath on enlistment to serve on if required in time of war, many of them proved unfaithful to this oath when their loyalty was put to the test.

It was a sad blow to the reputation of the Battalion, and more especially to the reputation of the Afridis. For many years the Battalion had been proud of its Afridi companies, and in 1901, when one company of Afridis was mustered out on reorganization, much regret was felt. Many of the remnant of loyal Afridis were first rate soldiers, and though they were not permitted to serve with the Regiment in Mesopotamia, they proved the excellence of their fighting qualities later on in East Africa, where they served with much distinction in the 40th Pathans. The Officer Commanding that Battalion, after the war, reported that he had never come across such a well-trained and efficient company as that company of Afridis. The Indian officers, Subadars Jan Gul and Sher Akhmad, and Jemadar Mirdast particularly distinguished themselves ; all earned a decoration.

From May to December, 1915, the Battalion served at Bannu and in the Tochi, and shared the duties and fatigues, alarms and excursions incidental to life on the North-West Frontier. During these months there were no large operations, and, therefore, no opportunity of distinction or of earning the war medal. The arduous nature of the duties may be partly gauged by the large number of detachments which the Battalion had to find. At one time it found no less than ten detachments varying from 2 British officers and 75 men at Idak, to a small post of 16 men at Mirzail; altogether 3 British officers, 8 Indian officers and 405 men were on detachment.

While serving in the Tochi orders were suddenly received to prepare for service overseas, and it was generally realized that the destination would be Mesopotamia.

The situation in that country was at this time considered very critical. After the Battle of Ctesiphon, General Townshend had been defeated and forced to retire to Kut, where, with the remnants of the 6th Division, he was besieged by the Turks. All available troops from India and the Indian Corps from France were being hastened to Mesopotamia. The North-West Frontier of India became for the time being of secondary importance; and the risk entailed by denuding it of troops had to be taken.

Mesopotamia On December 30th, 1915, the Battalion embarked at Karachi under the command of Lieutenant-Colonel L. B. Walton, with a strength of 13 British officers, 18 Indian officers, 713 Indian other ranks, and 45 followers.

1916 It arrived, after a calm voyage, at Basrah on January 3rd, but did not disembark till the following day. Thus, by a few days, it had the misfortune to miss earning the 1915 Star. After experiencing much discomfort, owing to a change of orders regarding the Battalion's location at Basrah, and owing to a heavy downpour of rain that same night, the Battalion left Basrah on January 10th with No. 6 Marching Echelon, in which the 1/6th Devons (Territorial) and the Sussex Battery R.F.A. (Territorial) were included. This echelon marched up along the Tigris, reaching Amarah on

Map Nº I.
Sketch Map of MESOPOTAMIA
Mosul
Arbil
Kurdistan
Altun Kupri
Sulaimaniyah
Lesser Zab
Kirkuk
Qalat as Sharqat
Diala R.
Sinneh
Fathah
Adhaim R.
Kifri
Zohab
PERSIA
Baiji
Qasr-i-Shirin
Sar-i-Pul
Pa-i-Taq
Hamadan
Tekrit
Jabal
Qarah Tappeh
Taq-i-Girrah
Khaniqin
Sarmil
Karind
Kirmanshah
R. Tigris
Mirjanah
Samarrah
RS.
Hamrin
Chashmeh
Sufid
Mahidasht
Kurdarrah
Balad
Shahroban
Abu Jisrah
Mendali
Baqubah
Buhriz
Balad Ruz
Luristan.
Hit
Euphrates R.
BAGHDAD
Jasen
Falujah
RS.
Hinaidi
Diala
Bawi
Ctesiphon
Pusht.

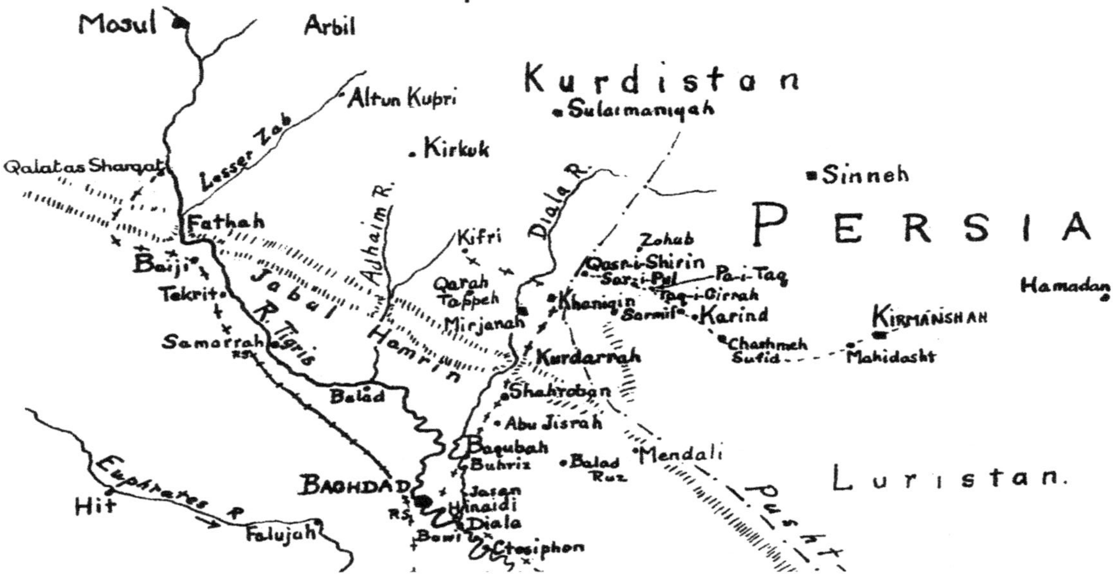

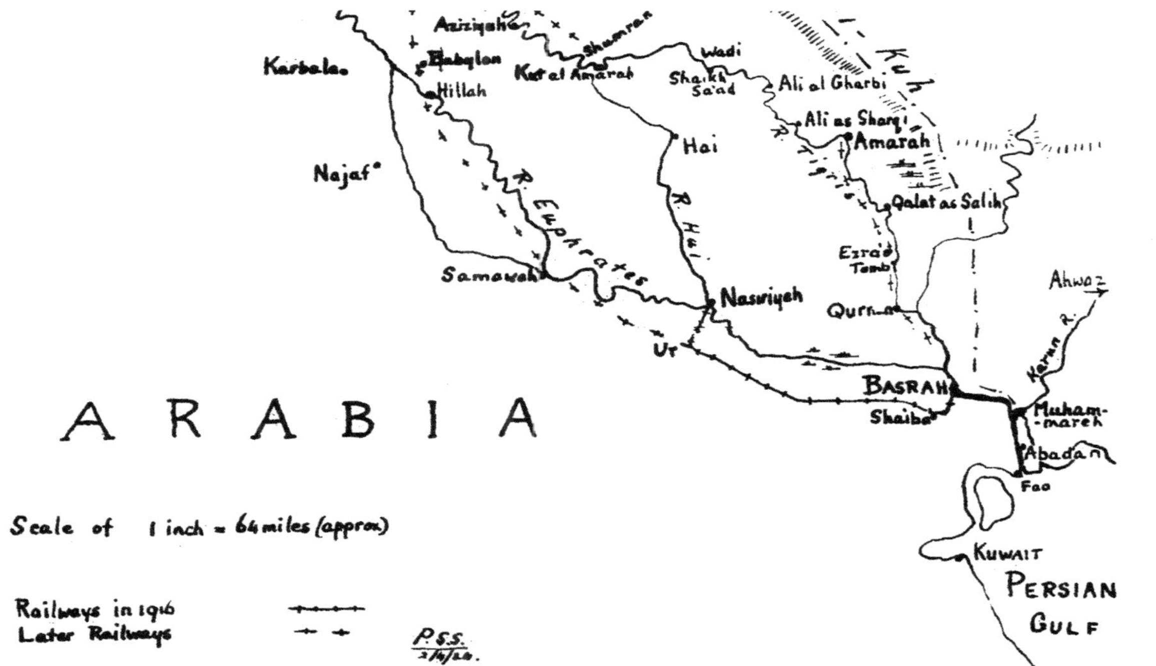

ARABIA
Karbala
Najaf
Azizyah
Babylon
Hillah
Shumran
Kut el Amarah
Wadi
Shaikh Sa'ad
Ali al Gharbi
Ali as Sharqi
Amarah
Qalat as Salih
Kuh
Hai
R. Hai
R. Euphrates
Samawah
Nasiriyah
Ur
Ezra's Tomb
Qurna
Ahwaz
Karun R.
BASRAH
Shaiba
Muham-mareh
Abadan
Fao
Kuwait
PERSIAN GULF
Scale of 1 inch = 64 miles (approx)
Railways in 1916
Later Railways
P.S.S.
2/4/20.

January 23rd, Ali al Gharbi on February 2nd, and Shaikh Sa'ad on February 5th. This march was rendered very trying by the vileness of the weather, which flooded the numerous small irrigation cuts and turned the road into a morass. Arab thieves also contributed to the discomfort by looting some of the Commanding Officer's kit.

The situation of the Tigris Corps and of the 6th Division in Kut necessitated the utmost speed in the despatch of reinforcements up-stream. Major-General F. Aylmer, V.C., now commanding the Mesopotamia Expeditionary Force in place of General Nixon, had recently beaten the Turks both at Shaikh Sa'ad on January 6th and 7th, and at Wadi on January 13th and 14th ; but neither battle had been decisive, and even after a third battle at Umm al Hannah on January 21st, the Turkish Army still lay between the relieving and beleaguered British forces. Meanwhile General Aylmer was busy organizing his reinforcements and preparing for a fresh attack on the Turks.

The 26th Punjabis remained at Shaikh Sa'ad on the right bank from February 5th to 17th, building and improving the defences of that place.

On February 18th the Regiment marched on up-stream to Orah, and joined the 36th Brigade under the command of Brigadier-General Christian, and on the following day took part in a reconnaissance round the enemy's left flank. On the night of February 21st/22nd it crossed the river, and, with the rest of the Brigade, bivouacked in the Sanna position. While holding this position it experienced, for the first time, bombing by hostile aeroplanes ; and the machine gunners realized what a difficult target an aeroplane offers. About a fortnight was spent here while preparations were being made for a further advance against the Turks. During this pause the weather was execrable, rendering all movement very difficult.

On March 7th, by which date the **Battle of Dujailah** weather had improved, plans had been completed by General Aylmer and General Gorringe (Chief of Staff) for a surprise attack on the Turks on the right bank of the Tigris. An attempt was to be made to work round his right flank and seizing the Dujailah position, to cut off his troops farther down-stream. The plans for getting into position under cover of darkness preparatory to an attack at daybreak, worked out almost like clockwork. (See Map 2.)

At 6.80 p.m. on March 7th the 36th Brigade, with the 26th Punjabis leading, left Sanna Camp for the rendezvous at " Ruined Hut," which was reached at 8.15 p.m. There the Brigade became part of Column " A " [36th Brigade, 8th Field Battery, R.F.A., 34th Pioneers (less four companies) and one company, S. and M.]. This column, with the 26th Punjabis again leading, left " Ruined Hut " at 10.25 p.m., and marching through the night by compass bearing, arrived at 6.80 a.m. at the appointed place in the Dujailah Depression, where it halted. The column had to carry out two changes of direction during this night march, a fact which enhances the credit of arriving correctly at its destination at the proper time.

It was not till 7 a.m. that the Turks first appeared to become aware of the presence of the column on their flank. About this time scattered bodies of Arab Cavalry were seen galloping away in a south-west direction from the Redoubt. These were fired on by the machine guns of the Regiment, and at the same time the artillery opened fire on the Turkish camp. But it was not till 9 a.m., two and a half hours after arriving at the position of deployment, that the infantry were ordered to advance. What caused this delay, and whether anyone was to blame for the apparent lack of initiative which permitted it, has been the subject of much comment, speculation. and enquiry. The regimental officer merely saw the result. Bodies of Turkish infantry were seen moving from the Redoubt in a north-westerly direction, apparently to occupy outlying trenches, and Turkish reinforcements were pushed into the Redoubt, so that instead of assaulting an almost unoccupied position by surprise at dawn, our infantry found themselves attacking a strongly-held and well-prepared position across open country in daylight.

The rôle of the 36th Brigade was to protect the left flank of the main attack on the Redoubt (the attack was to be made by the 28th and 9th Brigades) by advancing on the Shrine of Imam al Mansur. The 36th Brigade moved out in two lines, with the Regiment on the left of the first line. The machine guns were brigaded and had orders to protect the left flank.

At 10 a.m. the 82nd Punjabis, on the right of the 26th, changed their line of direction, and joined in with the main attack. No. 3 Double Company conformed to this move, but the remainder of the Brigade continued in the original direction.

No. 3 Double Company reached a line about 800 yards from the Turkish trenches, where they were held up by heavy rifle and machine-gun fire, and where Major I. M. Little, commanding the double company, was badly wounded. This line was held till about 2.15 p.m., when the officer in command of the double company of the other regiment on the right, reported that he had run out of ammunition and intended to retire. No. 3 Double Company was itself also rather short of ammunition, and was being outflanked by the Turks on the left. Accordingly it also retired, and reached the Depression at about 4.30 p.m. An hour or so later it moved forward and joined up with the remainder of the 26th again.

Meanwhile the 26th had gone on with the 36th Brigade and, coming under fairly heavy machine-gun fire, had suffered a few casualties. At about 11.30 a.m. Turkish cavalry were located on the left flank, and the Brigade halted and threw back its left flank to deal with them ; but the threatened attack did not materialize, and at 1.40 p.m. the Brigade was ordered to change direction, and attack the Redoubt in support of the 28th Brigade.

The Brigade moved to the attack with the 1/6th Devons in the first line, and the 26th and 100 rifles of the 62nd Punjabis in the second line. At about 1,500 yards from the Redoubt the Regiment came under heavy rifle and machine-gun fire from the front and right front. This continued until the advance had reached a point 800 yards south of the Redoubt, where the fire reached its greatest intensity, and caused most of the casualties.

Two hundred yards farther on the advance of the Regiment was held up by belts of fire from the enemy's machine guns, and from trenches running north-east and south-west on the left front. A few isolated groups managed to push on a little farther, with one of which was Lieutenant-Colonel Thompson, commanding No. 1 Double Company, who made the following observations :—

(1) Shells from our guns were bursting beyond the trenches to the south of the Redoubt, and these were the trenches which were holding up the advance.

(2) No artillery fire seemed to be directed against the enemy's machine guns.

(3) Without reinforcements no further advance was possible.

(4) The Turkish trenches were most skilfully concealed, and throughout the advance could be rarely seen.

During this phase of the battle the regimental machine guns, under Second-Lieutenant Farwell, were man-handled into action on the left of the Regiment, and were engaged in protecting the left flank. Until about 5.30 p.m. only a few poor targets were obtained. At that hour they opened fire on a large body of Arab cavalry and inflicted considerable loss.

At dusk the Regiment withdrew and **After the** was reformed in a large dry irrigation **Battle** canal, 1,000 yards in rear of the firing-line. The wounded were also collected there by the regimental stretcher-bearers. The latter worked very well under Captain Malcolmson, Indian Medical Service, and despite the darkness and continual sniping, brought in all the wounded except six, two of whom were known to have been seriously wounded. Parties of Arabs came right among the troops, and seriously hampered their work. These Arabs tried to lead parties astray by showing lights, and, if surprised by a body of troops, they feigned death. Captain Malcolmson actually captured one Arab doing this.

The attempt to capture the Dujailah Redoubt by a surprise attack had failed, and it was found to be impossible to complete a deliberate attack that day. Unfortunately, owing to the interposition of the Turkish troops between our forces and the river, there was no means of obtaining sufficient water to hold on where they were, and renew the attack next day. The British force was therefore obliged to withdraw again all the way back to the place where it had started. This was both disheartening and difficult.

Orders for the withdrawal were received **Withdrawal** about midnight. At 2 a.m. the 36th Brigade commenced to withdraw, and by daybreak had taken up and lightly entrenched a position in the Dujailah Depression to cover the evacuation of the wounded and the retirement of the main body. During the morning about 150 men who had become separated from their units were able to fall back to this position, our rifle and machine-gun fire driving back the swarms of Arabs who were trying to cut them off.

At 10.30 a.m. the 36th Brigade, acting as rearguard to the column, commenced the main retirement. The only casualties sustained were a few from shell-fire, as the Turks made no serious attempt to interfere with the retirement. The lack of water was very severely felt. Since leaving the Tigris on the 7th, till late on the 9th, the only water obtained was a very little brackish water from some wells in the depression. The Regiment eventually reached camp on the left bank of the Tigris at Orah at about midnight of March 9th and 10th.

Casualties The casualties of the 26th during this operation were :—

Killed and missing, 13 Indian other ranks.

Wounded, 3 British officers, 3 Indian officers, 85 Indian other ranks, and 1 follower.

Major Lawson Major O. H. Lawson, who had been dangerously wounded in the stomach on March 8th, and who had to be evacuated part of the way in an army transport cart, no ambulance being available, unfortunately succumbed to his wound at Amarah on the 11th.

His loss was greatly deplored by all ranks. He had spent his whole service in the Regiment, and by his sterling qualities had endeared himself to officers and men. He worked hard, but always most unostentatiously. For many years he performed the onerous work of Quartermaster, and spent what little spare time he had in training the signallers. To him was in a great measure due the efficiency of the Regiment in signalling. The officers specially had cause to be grateful to him for his constant and profitable care of the mess.

General Situation Thus ended what seemed a most promising operation for the relief of the beleaguered garrison at Kut. Great must have been the disappointment both in that garrison and in the relieving force at its failure. General Aylmer afterwards continued to hammer at the Turks with what troops he could muster, but he never again came so near to breaking through. The capture of the Turkish Hannah—Falahiyah position on

April 5th, the assault on the Sunnaiyat position on April 9th, the capture of Beit Aiessa on the 17th, with the subsequent heavy counter-attacks by the Turks, these and other later attacks did much towards wearing down Turkish resistance ; but the British relieving force did not again get as near to its objective as it did on March 8th.

After the Dujailah battle the Regiment was employed for the next fortnight in digging flood *bands* to keep out the river, which began to rise very high. On the night of March 27th and 28th the camp was sniped, and on the following morning a small party under Captain Cargill went out and succeeded in ambushing a party of mounted Arabs, wounding 3 and killing a horse.

The 36th Brigade, being corps troops, and attached to no division, was, during the ensuing month, used as a reserve whenever required, and was continually being moved from one part of the line to another. On March 30th the Brigade moved its camp from the left to the right bank.

At daybreak on April 5th the Turkish position at Hannah was attacked by the 13th Division under cover of an artillery bombardment from both banks of the river, and under cover of overhead indirect machine-gun fire. This latter was provided by the brigaded machine guns of the 36th Brigade, in which the 26th machine guns were included. After the attack had been successfully pushed through, the machine gun section rejoined the Regiment at the Sanna position, to hold which the Regiment had been pushed up.

On April 13th the Regiment rejoined the rest of the Brigade at Sandy Ridge. On 17th/18th at midnight, after the battle of Beit Aiessa, the Brigade was hurriedly sent up to Abu Roman in support of the 8th Brigade, which had been heavily counter-attacked in the Beit Aiessa position by the Turks. The following morning the Brigade marched to the " Triangle " and took over trenches from the 37th and 38th Brigades. On the 19th it marched back to Abu Roman, and at 10.30 p.m. on 21st crossed to the left bank of the Tigris to take part in the operation known as the Second Attack on Sunnaiyat.

At 1 a.m. on the 22nd the Brigade arrived **Sunnaiyat** in position supporting the 7th Division, which was to carry out the assault, and was ordered to move up to support the 21st Brigade. It moved up in two lines, with the 26th Punjabis on the right of the first line, and, after moving forward about 1,000 yards, was ordered to occupy a line of trenches. Owing to the flooded state of the ground the assault was unsuccessful, and therefore the Battalion remained all day in these trenches. Only nine casualties were sustained by the Regiment, chiefly from long-range rifle fire. The regimental machine guns were in action with a massed machine-gun battery on the right bank, supporting this operation. At 8 p.m. the Brigade re-crossed the river and marched back to Abu Roman.

On April 24th the Brigade supported an attack by the 9th Brigade, and after being relieved by the 8th Brigade marched back to the north-east of the "Triangle" and bivouacked. On the 29th the Brigade took over the first line trenches south of the "Triangle."

On April 30th at 4.30 p.m. the startling **Fall of Kut** news was received that Kut had surrendered the previous day, and that a three days' armistice had been arranged.

This news was very depressing to the moral of the troops. All ranks had been striving day after day to relieve the beleaguered troops under General Townshend, and in spite of the stubborn fighting of the Turks their efforts would probably in time have proved successful. The season, however, was unfavourable. Floods on both sides of the Tigris hampered all movements of the relieving force, and eventually brought the force to a standstill.

The gallant garrison of Kut maintained its position as long as food was available. The troops existed many months on reduced rations, and all the survivors were much emaciated by want of proper nourishment. Large numbers of the garrison had already succumbed to disease, induced or aggravated by hardship and starvation.

It was only the complete exhaustion of all provisions, and the hopelessness of immediate relief, which forced them to surrender. Before doing so they destroyed all their guns, rifles, and ammunition, so that they should not fall into the hands of the Turks.

After a siege of 143 days the garrison was reduced by about one third to some 7,000 all told. With the exception of a few seriously ill or wounded, these were all taken prisoner by the Turks, and suffered further unparalleled hardships at the hands of their cruel captors before the remnants obtained their release at the end of the war.

Summer, 1916 After the fall of Kut there commenced for the Tigris Force a most trying hot weather period. Moral, depressed by want of success, was further reduced by climate and unhealthy conditions, so that disease soon began to take heavy toll.

The Turk proved more or less inactive, and in fact, withdrew here and there, so that the Tigris Force gradually took over several of the positions which the Turks had recently defended with much tenacity.

On May 19th it was reported that the Turks had evacuated their Beit Aiessa position, and the 36th Brigade moved forward at short notice that day. On the 21st the 9th Brigade occupied Dujailah without encountering any resistance, and the remainder of the force marched to Imam al Mansur, arriving at 2.30 p.m. The day was very hot, and great difficulty was experienced in obtaining water for the troops. A certain amount was brought in motor lorries from Abu Roman, and after arriving at Imam al Mansur, parties were sent down to the Tigris at Magasis. These parties were shelled by the Turks, and suffered a few casualties.

At 4 a.m. the following morning the Brigade marched to Magasis Canal, and took up an outpost position stretching from the Tigris along the Magasis Canal to Dujailah. Enemy aeroplanes showed considerable activity during the next few days, dropping bombs almost daily, but they caused comparatively little damage. During this period the Turk had almost complete command of the air, as the few serviceable machines on our side were not capable of fighting the German Fokkers.

During June and July existence was rather dull and monotonous. Brigades relieved each other in the outpost line at regular intervals, and no operations of importance took place. On June 14th a sniping picquet of ours fired on some Turks, relieving a picquet on the farther bank, and seemed to cause some casualties. On the night of July 2nd/3rd Arab marauders attacked two of our picquets with bombs and rifle fire, and the picquets replied. On subsequent nights ambushes were laid for them, but without success.

The weather was very hot, and water not too plentiful. What there was was obtained from wells dug in the Dujailah depression. The men felt the scarcity of water severely, particularly for washing. Sickness, both among British officers and Indian ranks was heavy, and there was practically no means of breaking the monotony. There was no equipment for football or hockey, so rounders was instituted, and was played most evenings. Dust storms and dust " devils " blew through the camp almost daily, making life still more unpleasant.

On July 2nd Lieutenant-Colonel L. B. Walton took over command of the 36th Brigade, and Lieutenant-Colonel I. F. R. Thompson became acting-commandant of the Battalion.

Brigadier-General Walton, which rank he now received, did not again revert to regimental duty. He continued to command the 36th Brigade for some months, but his health had been undermined, and he had to be invalided to India, where he died in the following year.

His geniality and sportsmanship had made him popular with all ranks. During his tenure of command the Battalion continued to excel in musketry and sports, as well as in military efficiency. His good service was recognized, not only by his promotion to command a brigade, but also by the award of the brevet rank of colonel.

At the end of July a pleasant change occurred. The Regiment relieved the 82nd Punjabis in the detached outpost of Imam al Mansur (an old Arab tomb), two companies going in advance on the 24th, and the remainder following next day. The change was much appreciated for several reasons. It broke the monotony; the well-water there was cool and good; and, being on detached outpost, life was considerably more interesting. The British officers had their mess and quarters inside the tomb, which was very much cooler than a tent. The regimental signallers were also quartered in the tomb, and had an observation post on top of it. The Regiment was distributed in three redoubts, dug round the tomb.

On August 1st the Turks appeared to celebrate the close of the Ramzan (fast) by sending their daily cavalry patrols very much nearer our lines than usual. These daily patrols used occasionally to be fired upon by our machine guns, or by the section of 18-pounders from No. 1 Redoubt. On the night of August 8th/9th Arab thieves tried to enter No. 1 Redoubt under cover of some fairly heavy sniping, but were not successful.

During August the weather became considerably cooler, as the " Shamal " started blowing. Rations,

E

however, were very poor, and the strength of the Regiment was so much depleted by sickness, principally scurvy, that a reinforcement of 155 rifles from the 62nd Punjabis had to be sent out to take over Nos. 2 and 3 Redoubts.

Increased Activity The approach of the cooler weather was marked by increased activity on both sides. Our cavalry made more reconnaissances; an observation post was erected, and 60-pounder gun-pits were dug in the Dujailah Depression near the Imam, the Regiment supplying the working parties; and barrages were registered by the artillery in front of our line. This activity may have been due partly to the weather, but partly also to the appointment of a new force commander, Sir Stanley Maude, who had till now been in command of the 13th Division. He replaced Sir Fenton Aylmer.

As regards the enemy, their cavalry patrols became more bold and active, several large convoys of sheep were observed moving up the Hai towards Kut, and every night for about a week towards the end of September we heard heavy firing from the Turkish lines, as if they were carrying out night firing practices.

On the evening of September 27th the Turks extensively bombarded the whole of the Sinn area. They fired about 25 high explosive shells at our position at Imam, but caused no casualties in the Regiment. From " duds " picked up, these were found to be 4.7 British naval shells, and were presumably fired from the captured monitor *Firefly*. Arab thieves were very active during September, and ambushes were frequently laid for them, but without success.

On October 2nd the Regiment was relieved in the Imam by the 3rd Brahmans, and concentrated with the remainder of the 36th Brigade in a camp, about 4,000 yards in rear. Battalion and Brigade Training was there carried out for about a fortnight. On October 16th the Brigade again took over the outpost line, 80 rifles of the Regiment reinforcing the 62nd Punjabis in the " Pentagon," the remainder of the Regiment being in local reserve.

On the 20th the Commander-in-Chief in India, Sir Charles Monro, in the course of a hurried tour while on his way out to India, came and saw all available British and Indian officers.

On the night of October 28th the Regiment together with the 1/4th Hampshire Regiment and the 128th Pioneers, dug a communication trench along the river bank from Magasis pumping station to Nasariyeh pumping station. The task had been reconnoitred the previous evening. Though there was considerable sniping from enemy picquets on the farther bank, no casualties were sustained.

Reorganiza- Under the new Army Commander much reorganization was being carried out, and **tion** preparations were pushed forward for a limited offensive against the Turks. The Tigris Force was divided into two Corps :—

Ist Indian Army Corps :
  3rd Division (7th, 8th and 9th Brigades).
  7th Division (19th, 21st and 28th Brigades).

IIIrd Indian Army Corps :
  13th Division, all British (38th, 39th and 40th Brigades.
  14th Division (35th, 36th and 37th Brigades).

The 26th Punjabis remained in the 36th Brigade as before.

On November 13th the Brigade concentrated in the central Sinn area camp for training. Enemy aeroplanes frequently came over and dropped bombs, but the only casualty they caused in the Regiment was one havildar wounded on November 17th. On November 23rd the Regiment took over the Dujailah Redoubt. The rest of the month was spent in improving the defences there, and in carrying on company training.

Winter Offensive

The weather by now had become much colder. Rain started early in December, and delayed the start of our offensive, which had been planned with the limited object of improving our position astride the Hai. It commenced eventually on December 12th. Prior to this Captain J. E. Shearer had been carrying out night patrols and reconnaissances towards the Turkish lines. For this special work he was officially thanked.

At 2.15 a.m. on December 14th, the Regiment took up an outpost position along the line of the old Turkish Redoubts. Regimental Headquarters remained in the Dujailah Redoubt.

The Regiment's rôle was actually left flank guard to our main operation against the Turkish position on the right bank of the Tigris opposite Kut. Nothing of importance happened here, except slight sniping one night against "C" and "D" Companies.

Hai Bridge-head

At 5.45 a.m. on the 18th the whole Regiment concentrated at Atab, whence it marched across the Hai to S.9 and bivouacked for the night.

At 6 a.m. on the 19th "B" Company moved out to take up a line of observation and at 9 a.m. the rest of the Regiment marched to take up an outpost line connecting with a brigade of the 13th Division on the right.

On the 20th large numbers of Arabs were observed along a *band* about 2,000 yards from our line and a few shots were exchanged with them. They remained there watching us, and occasionally sniping us, the whole time we were holding this outpost line. The weather was very bad : cold with a great deal of heavy rain. The low-lying ground in rear of our line quickly became a marsh, from which a duck was occasionally obtained for the mess.

On the evening of Christmas Day a report was received that the enemy were digging in on a line near our outposts, and we had to send three consecutive patrols including an officer's patrol, to confirm this. Each patrol brought a negative report.

During the period the Regiment was occupying this outpost line, a new strong point line was being dug some way to the rear as a protection to the Hai Bridge. On the 28th the Regiment received orders to fill in all their existing trenches, and then move to a camp behind the new strong point line. The move was postponed owing to heavy rain, but it took place the next night (29/30th). Great difficulty was experienced with the transport, owing to the darkness, heavy rain, and sodden ground. The screen left to cover the withdrawal retired at 8 a.m., and as it was doing so a Turkish deserter came in and gave himself up.

1917

Abdul Hassan Bend

The Regiment then went into camp with the remainder of the Brigade at Nahr Bassouyah, a fairly comfortable camp, where it stayed over the New Year and until the heavy fighting at the end of January. The 3rd Lahore Division had been brought across the Tigris and was attacking the Turkish trenches in the Abdul Hassan bend between the Tigris and the east bank of the Ha. It was realized that as soon as the Turks had been driven out of this bend, the IIIrd Corps would have to continue the same task to the west of the Hail. The capture of the Abdul Hassan bend was actually completed between January 9th and 19th.

During this comparatively comfortable pause, a night operation was carried out which was anything but comfortable. On January 2nd orders were received that the Brigade would move out that night and blow up a tower some 2,000 yards in front of our late outpost position. This necessitated an eighteen-mile march there and back. Except that there was no fighting, it would be difficult to imagine a worse night. Starting at 6.30 p.m., it took five hours to reach the objective, and another one and a half hours for the Pioneers to destroy the tower. The night was bitterly cold, with a gale of wind blowing from the east, and, to make matters worse, at 1.30 a.m., it started to rain, and later this turned to sleet. It was 7.30 a.m. before the Regiment marched back tired, and nearly frozen, into camp.

On January 8th another night march took place, which, though constituting a tiring and anxious operation lasting fifteen hours, was not as bad as the night march of the 2nd/3rd. Moving out at 11.25 p.m. we arrived at the Divisional rendezvous at 11.55 p.m., and at the position of deployment at 4.30 a.m. There, after deploying, we halted till 5.30 a.m., when the Brigade advanced on a bearing of 12° in two lines of battalions at 100 yards distance. Each battalion was in two lines of companies in fours. We were the leading battalion on the right, the 82nd being on our left.

After advancing 3,000 yards we halted at 6.10 a.m. At 6.25 a.m. orders were received to halt and wait further orders, and here we learnt for the first time the object of the night march and subsequent advance.

It appeared that the 3rd Lahore Division were to attack and take some strong enemy trenches in the Abdul Hassan bend that morning, and we were to make a strong demonstration against the Shumran bend in the hope of drawing off the enemy's reserves from the real object of the attack.

By 8.30 a.m. the mist showed no sign of clearing, and it was still impossible to see more than a few yards. In these circumstances it was of course no use attempting to make a demonstration, and at 9 a.m. we received orders that the force would return if the mist had not cleared by 10.30. The next hour and a half was an anxious time, as it cannot be said that anyone was desirous of advancing farther. A demonstration in front of a vigilant enemy well armed with machine guns is an unpleasant duty at any time, and over a dead-flat and featureless country, even more so. than ordinarily. Further, it was by no means certain how far we were from the enemy's trenches, and it was quite possible that the mist might suddenly clear and discover us well within range.

However, at 10.30 the mist was still heavy, and orders were received to return to camp, which we reached at 2.30 p.m.

For the next fortnight we were kept hard at work both by day and night, digging communication and other trenches in the Hai bridge-head. During this time many rumours were circulating as to what the coming operations would be, but only one prospect seemed certain, and that was that there must be some heavy fighting in front of us.

By January 20th the 3rd Division had finally cleared the Turks from the Abdul Hassan bend, and it was now our task to clear them from the other bank of the Hai.

On January 24th we at last received **Dahra Bend** our orders, and these showed that the 36th **Operations** Brigade were to be corps reserve ; the 26th Punjabis to be in a position of readiness by 9 a.m. the next day. The remainder of the Brigade was to stay in camp.

At 8.30 a.m. on the 25th the Regiment marched to its position of readiness—the bridge over the Hai—and sat down to wait. We knew that the 13th Division was to attack the enemy's trenches on the west of the Hai, and news of the result of the fighting was eagerly awaited.

At first the news that filtered through was good, but about 1.30 p.m. rumours reached us that all was not going as well as might be, and at 2 p.m. orders were received for the Regiment to move to a point just north of the junction of the main road and Nahr Bassouyah—known as " Oxford Circus "—and there join up with the remainder of the Brigade. We arrived there at 2.30 p.m., and, joining the rest of the Brigade, were ordered to move on and take over the strong points in rear of the 39th Brigade (13th Division) trenches.

To do this we had to cross about 1,000 yards of open ground to a nullah which led up to the strong points. The leading company had got halfway across when the Turks opened fairly heavy fire with all sorts and conditions of guns, from 5.9 howitzers to small camel guns. Although the Turks had our exact range, and must, in the clear evening light, have been able to see us distinctly, their fire, as was afterwards usually found to be the case, was singularly ineffective, and we only had 3 men and 1 mule wounded.

Arrived in our position we heard the story of the day's happenings. How the 39th Brigade having successfully taken the Turkish front-line trench, and how later, owing chiefly to being too tightly packed in the trenches, they had been bombed out and forced to retire to the advanced trenches, from which they had originally attacked.

About 5.30 p.m. orders were received that the Regiment, in conjunction with the 62nd Punjabis, were to attack the enemy's trenches from which the 39th had been forced to retire, and re-take them before 7 p.m.

It was already growing dark; no one in the two battalions, or for that matter in the whole brigade, had ever been on the ground before, or even knew the positions of our front-line trenches; and a preliminary reconnaissance was out of the question. It can be imagined, therefore, with what feelings of dismay the order was received.

Happily the consternation caused by this order was short lived, for half an hour later the order was cancelled.

No further orders were received till 10 p.m., when the 22nd Punjabis and ourselves were told to take over the three front-line trenches from the 39th

Brigade ; the move to be completed by 10 a.m. the next morning.

We started to move up immediately, the 82nd taking over the two front-line trenches, and we, the third, known as "Emperor's" trench. Owing to the darkness, and the fact that we had never been in trenches before, it was 1.30 a.m. before we had completed the move. The night was bitterly cold, and both officers and men found, owing to the lack of blankets or other covering, that sleep was, if not impossible, at least difficult.

Attack of January 26th On that cold early morning at 3 a.m. orders were received that the 82nd Punjabis were to assault the enemy's trenches from P.15 to P. 12 at 9.40 a.m., and that we were to detail 30 bombers to go with them to secure their right flank by bombing up from P.15 and blocking the communication trench leading up from that point.

The remainder of the disturbed night passed uneventfully. At 8 a.m. instructions were received that we were to hold ourselves in readiness to support the 82nd Punjabis should they be heavily counter-attacked. At 9 a.m. operations were postponed for an hour, and we received orders to furnish nearly 200 men to move ammunition up to the front-line trench. Owing to this demand on our strength we found ourselves holding "Emperor's" trench, which was over 1,000 yards long, with less than 500 men, consequently platoons were strung out at wide intervals and on long fronts.

At 10.30 a.m. our guns, which had been shelling the enemy's first three lines of trenches for the last fifty minutes, began their intense bombardment. It appeared to us, who had never before seen rapid gun-fire at such close quarters, that everything in the neighbourhood of where the shells were falling must be totally destroyed. From where we were we could see little or nothing of what was happening in front. We ourselves were being subjected to the concentrated fire of all the Turkish guns, but thanks to the depth and stoutness of our trenches received but little damage.

At 10.40 a.m. the 82nd Punjabis, together with our bombers on their right, went " over the top," and sustaining but few casualties, arrived at the Turkish front-line trench to find it deserted, except for dead and wounded Turks. It then came to our turn to cross the open. Orders were received at 11.5 a.m. to reinforce the 82nd Punjabis, but owing to the length and depth of the trench and the number of traverses, the communicating of orders was difficult, it was not till 11.85 that the advance commenced.

Lieutenant-Colonel Thompson, who was in command, gave orders that the Regiment should advance in four lines, each of four platoons, and that the left platoon should direct, " A " and " C " Companies forming the first two lines, and " B " and " D " the next two, regimental Headquarters going with " A " Company.

The fact that both flanks of " Emperor's " trench were slightly refused caused the company on the left (" A ") to face half-left, and that on the right (" D ") to face half-right. This was not realized at the time. When " A " and " C " Companies went over the top, " A " Company, under Captain Shearer, failed to get their right direction, and went too much to the left. " C " Company, under Lieutenant Farwell, being in the centre, and so facing their proper front, arrived almost directly at P.12A, which was the left of the line held by the 82nd Punjabis. " B " Company, under Second-Lieutenant Hunt, followed " A " Company, and eventually joined up with it. " D " Company on the right, under Major Maude, went over straight to its front, and arrived near P.15 on the right of the 82nd Punjabis.

" A " and " B " Companies and Headquarters, keeping too much to the left, finally reached the enemy's trench at P.16, which was the Turkish very-much-refused right flank. To reach it they had to cross about 1,500 yards of open, and suffered heavily from shell fire. It was just as Headquarters arrived at the enemy's trench that the Commanding Officer, Lieutenant-Colonel J. F. R. Thompson, was unfortunately killed.

Lance-Naik Channan Singh (signaller) bravely went back to the support trenches to bring up the Second-in-Command, Major Ivens, who in accordance with orders had been obliged to remain in rear with the " first reserve." Fire was heavy across the open and while Major Ivens and Channan Singh were crossing again, Major Ivens himself was wounded. For his gallant behaviour Channan Singh was awarded the Indian Distinguished Service Medal.

The command of the Regiment now devolved on Major Maude, but he was away on the right flank. The companies on the left (" A " and " C ") consolidated the trench they had captured, and sent out patrols to the right to get into touch with " C " Company. While doing so they were counter-attacked by the Turks, but succeeded in repulsing them. Ammunition then began to run short, so having gained touch with " C " Company, the two companies, " A " and " B " withdrew along the trench and joined up with " C " Company. This was at 2.15 p.m.

" C " and " D " Companies, in their advance across the open had also lost heavily from rifle and machine-gun fire, and all the while they were consolidating their position they were subjected to heavy shelling. The part of this trench, which had been assaulted and held for a few hours by the 39th Brigade on the previous day, was very badly knocked about by both our own and the Turkish bombardment. Moreover, it was narrow to start with, and became very congested. This made control of fire and consolidation very difficult.

The Regiment remained in the trenches all day, and consolidated them. In the evening, at 7.30 p.m., they were relieved by the 1/4th Hampshire Regiment and withdrew to " Emperor's " trench.

Thus ended the fighting of January 26th. The Regiment had carried out its task nobly. The men behaved splendidly, and earned high commendation. Even away in the 7th Division on the other side of the Tigris, their reputation spread. On January 30th the General Officer Commanding the 7th Division wrote to Colonel Thompson, of whose death he had not heard, saying : " I hear great deeds of the gallant 26th, and I knew they would always do well."

Unfortunately our casualties were rather heavy :—

Killed.—Lieutenant-Colonel Thompson, 2 Indian officers, and 24 Indian other ranks.

Wounded.—Major Ivens and Lieutenant Balkrishna, Indian Medical Service, 3 Indian officers, and 181 Indian other ranks.

Wounded and missing.—11 Indian other ranks.

Missing.—2 Indian other ranks.

Total casualties, 226.

Lieutenant-Colonel Thompson's death was particularly unfortunate. He was a very keen and efficient officer, and was much respected in the Regiment. His loss at a critical moment left the Regiment without any officer of long experience, and might have led to disastrous results. During his service he always set an example of patient industry and thoroughness in his work, which it would be difficult to surpass. Being a good rifle-shot himself, he did much to encourage good shooting in the Regiment. It is pleasing, therefore, when recording this to his memory, to be able to welcome his son into the Regiment to follow his example.

Completing Dahra Bend Operations After this for some days the Regiment did not take any very important part in the operations, which were still being continued, for the capture of the Dahra bend. The Turk was gradually driven back, and was subjected to relentless pressure, which wore down his resistance and greatly reduced his fighting strength.

On the evening of the 28th the Regiment was relieved in the support trenches by the 36th Sikhs, and went into camp with the 82nd Punjabis in rear. Here it remained in a position of readiness till 8.30 a.m. on February 4th, when it moved up into the support trenches again. Four men were killed, and three wounded by shell-fire on this occasion. On the night of February 5th, while the Regiment was working on communication trenches, Second - Lieutenant J. H. W. Stevenson was mortally wounded.

On February 6th the Regiment took over the picquets now on the line N.41—N.41D.—N.29C. (Map No. 3) from the 1/4th Hants under a fair amount of shelling, and on the following day still more picquets up to N.20G, from the 82nd Punjabis. " C " and " D " Companies held the picquet line, with " A " and " B " in reserve. During the night of the 7th/8th new picquets were dug by " C " and " D " Companies, with a view to establishing a complete line from N.20G to N.42. As there was a bright moon, and the enemy sniped continuously from a range of 250 yards, this work was very trying. By daybreak, however, they succeeded, with the loss of 3 Indian other ranks killed and 6 wounded, in establishing all picquets firmly.

During the 8th and the night of the 8th/9th work was continued on the new line and communication trenches. On February 9th the 13th Division, who were holding the continuation of our line on the left, carried out an attack, and were supported by the Regiment with rifle and machine-gun fire on suitable targets. Our casualties that day were 1 Indian other rank killed, and 1 Indian officer and 36 Indian other ranks wounded.

That night the Regiment was relieved by the 1/5th Buffs (13th Division), and withdrew to the support trenches, where it remained for the next five days, working on roads and communication trenches.

On February 15th it moved up in support of the 13th Division, who successfully attacked the Turks in the forenoon. In the afternoon it moved up still further in support of the 35th Brigade, who attacked at 1.30 p.m. These two attacks, though costly, were very successful, and were the culminating point of the Dahra bend phase of the operations. That evening we saw the very pleasing sight of large numbers of Turks coming across to meet our troops with their hands up. All the Turks remaining on the right bank of the Tigris, including their General Officer in Command, now surrendered.

General Situation General Maude had now accomplished the task which had been set him for the winter campaign. He had secured for the British an advantageous position astride the Hai, and could now prevent the Turks advancing either along that line or down the Tigris.

In doing so he had not incurred heavy casualties, and had no need to ask for reinforcements, which could be ill-spared from other more important theatres of war. He had, moreover, somewhat to his own surprise, unexpectedly inflicted far greater losses on the Turk, so that the latter was much depleted in numbers, and was in rather a precarious position.

On ascertaining this, General Maude asked for, and was granted, permission to press his advantage still further. He therefore planned the operations which began with the crossing at Shumran and eventually developed into the capture of Baghdad.

To deceive the enemy he sent the 7th Division to attack the Sannaiyat position. This was done on February 17th, but on that occasion unsuccessfully. On February 22nd complete success was achieved, and the position stormed and held.

Shumran Crossing Meanwhile the 26th Punjabis were with the force preparing to cross the Tigris at Shumran above Kut. On the morning of February 23rd the Regiment, with the rest

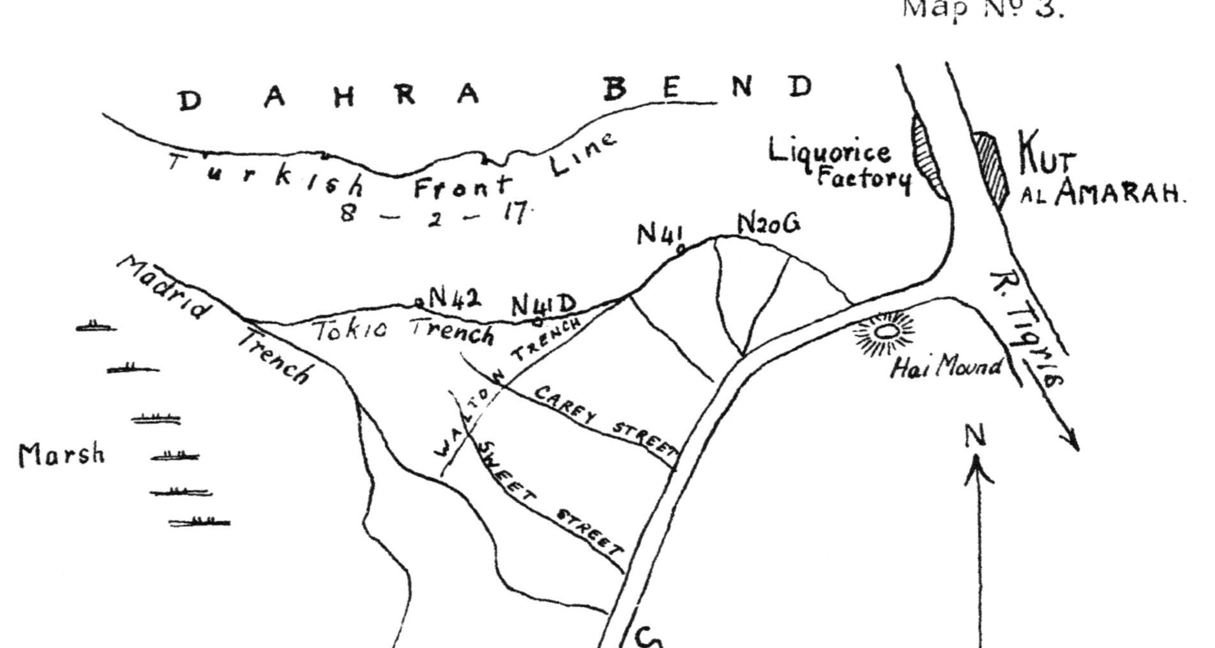

Map No 3.
DAHRA BEND
Turkish Front Line
8 — 2 — 17.
Liquorice Factory
Kut al AMARAH.
R. Tigris
N41
N20G
N42
N41D
N Trench
Tokio Trench
Madrid Trench
Walto
Carey Street
Sweet Street
Hai Mound
Marsh
N
S.

P.16.
Turkish
Position
on
26-1-24
P.12A
P.15
Bignell Trench
Thomson Trench
Emperor's Trench
Harvey Road
hatt al Hai
Kala Haji Fahan
S7
O S.9
O S.9
O S.9
Sketch Map
of
HAI BRIDGE-HEAD
Operations.
Scale: 3 inches = 1 mile.
P.S.S.
4/4/24

of the 36th Brigade, moved to a position of readiness for this crossing. The crossing was forced by the 37th Brigade, who threw three ferries across the river in the early morning of that day.

The Regiment's original orders were to be ready to move up and cross by the ferry at K.55 (the lowest) as soon as the 1/2nd Gurkhas had crossed by it. This ferry, however, was found to be very costly in casualties, owing to the heavy machine-gun and shell-fire to which it was exposed. Similarly M.30 ferry had to be abandoned, and eventually, at 7.30 p.m., the Regiment crossed with the rest of the Brigade by the bridge, which was situated at the extreme point of the bend, and which had been cleverly and rapidly thrown across the broad river by the Sappers. This sudden crossing of the river in the face of some opposition was a wonderful feat of organization and endurance.

By the time the Regiment crossed there was no shelling, but some rifle shots were flying overhead, and one man was wounded. The Regiment, now under the command of Major Ivens, who had partially recovered from his wound, spent the night behind the old Turkish breast-works on the river bank. At about midnight orders were received for the 36th Brigade to attack the left of the Turkish position across the neck of the Shumran bend at dawn. While moving at night Major Ivens was accidentally injured by falling into a dug-out in the darkness, and Major Maude took over command of the Regiment.

At 6.15 a.m. on February 24th the 36th Brigade attacked in two lines of battalions, the Regiment on the right of the second line. The attack was preceded by a five-minutes' intense artillery bombardment.

Shumran Bend The distance of the main Turkish position, which was along the line of hillocks across the neck of the bend, was about 1,500 yards.

There was, however, a minor intermediate position in some disused gun-pits about half-way across. " A " Company came to close quarters with a small party of Turks at this latter position, and overcame them. As the Battalion advanced further heavy rifle and machine-gun fire was encountered from some barracks on the right front, but a rapid deployment from artillery formation into extended order minimized casualties.

Eventually the Turkish position was successfully captured and consolidated. " D " Company, under Lieutenant Farwell, was pushed out to the right to clear out the barracks and gain touch with the river. The Turks were found to be still holding this position, and they opened a heavy burst of machine-gun fire. Soon, however, they started surrendering in batches, and the remnants were driven out into the open by accurate artillery fire, which had been called for. Two machine guns were captured by this company, and have since been presented to the Regiment as trophies.

" A " Company, under Captain Shearer, who were in the captured position, observed two Turkish field-guns, which had been deserted by their gunners, lying a short distance in front. In spite of heavy rifle fire they managed to reach these guns and bring them back into our lines, but Captain Shearer was wounded while doing so. For his gallantry on this occasion he was awarded the M.C. One of the two guns captured was found to be a British 18-pounder which the Turks had captured at Ahwaz, the other a modern Krupp field gun. In memory of this exploit the Regiment has been presented with a Turkish field-gun, which remains as a trophy at the regimental quarter-guard.

Having gained touch with the river and cleared the Turks from the vicinity, the Regiment consolidated its position and carried out salvage work, finally bivouacking in the captured position.

Thus finished a memorable day for the Regiment. The men acquitted themselves remarkably well, and

THREE COMMANDANTS WHO DO NOT APPEAR IN GROUPS.

Col. L. Dening, D.S.O.,
1894-1899

Col. A. A. E. Campbell,
1906-1910

Col. A. D. Cox,
1917-1921

were well led. The captures by the Regiment amounted to :—

>2 field guns.
>3 machine guns,
>500 prisoners.
>600 rifles, and a large quantity of ammunition and equipment.

Many were the rewards distributed in recognition of the Regiment's gallant conduct, including the D.S.O. given to Major Maude, who was in command.

Casualties, unfortunately, were also fairly heavy. One Indian officer and 10 Indian other ranks were killed ; 1 British officer, 1 Indian officer, and 75 Indian other ranks wounded.

General Results The results of this battle were far-reaching. The Turks, with their line of retreat threatened, retired helter-skelter from their positions in Kut and below that place, and not only abandoned large quantities of ammunition, equipment, and stores, but also surrendered in great numbers. Kut was recaptured by the British, but it was not till several days later that British troops actually entered Kut town, and then only in order to satisfy the insistent demands of the cinematograph photographer.

The whole Turkish Army was now in full retreat to Baghdad. Our cavalry, hampered by lack of mobility, the cause of which is much disputed, failed to seize what appeared a glorious opportunity. The whole British force, however, followed hard on the Turkish heels, and the 26th Punjabis took part with the rest of the 14th Division in this pursuit.

Advance to Baghdad Some of the marches were very long, and all were very dusty, and there was a constant scarcity of water. At the beginning of March the heat was increasing rapidly. Moral, however, was excellent, as all ranks were delighted, after a year of trench warfare, to feel

F

that they were part of a mobile victorious army with the Turks in full retreat in front of them. The longest and hottest of these marches was the one to Aziziyah, and the Turks themselves must have found it even more trying than we did, for the road was strewn with abandoned guns, ammunition, and stores.

On arrival at Bawi on the 7th it was found that the Turks were holding the line of the Diala, in a last vain effort to save Baghdad. On the 8th the 36th Brigade, with a battery of Royal Field Artillery, some Sappers and Miners, and two pontoons, marched from Bawi up the left bank of the Diala to demonstrate, and, if found to be practicable, to endeavour to force a crossing. The force reached a point on the Jasan road, about 5,000 yards from the Diala, and on advancing towards the river found it to be held in some strength. After being shelled and having suffered some casualties, it withdrew at dusk a short distance and bivouacked. Several officers' patrols were sent out, which confirmed the fact that the Turks were holding the river in strength.

The next day the force marched back to Bawi, and on starting were again shelled by the enemy. It was a long, hot march, and the lack of water was very much felt. On arrival near Bawi the Regiment was met by the ever-ready Quartermaster (Lieutenant Warren) with a couple of very welcome water carts, and the still more welcome news of plenty of hot tea and food waiting for the men in camp.

It turned out that our task had been purely a demonstration in order to prevent the Turks concentrating the whole of their force near the mouth of the Diala, where the main crossing was forced by the 13th Division. We were informed that the object of the demonstration had been completely attained.

On the same day as we marched away from Bawi to the north-east, a bridge had been thrown across the Tigris at Bawi, and the Cavalry Brigade and Ist Indian Army Corps had crossed and pushed on up the right bank. Baghdad railway station was occupied on the 10th, and on the 11th Baghdad city itself was seized simultaneously on both banks by the Ist Corps on the right, and the IIIrd Corps on the left bank. This crowning success by the British forces was acclaimed with delight by many of the inhabitants of Baghdad, who for centuries had lain under the blighting tyranny of the Turk, and had sunk from prosperity to extreme poverty.

The 36th Brigade bivouacked on the 11th and 12th at Hinaidi, south of Baghdad, and while there received orders to act as garrison of Baghdad. The 26th, under Major Maude, marched from Hinaidi to Baghdad on the 14th, and crossing the boat-bridge went into camp in a grove of palm trees, and took over duties on the right bank. Duties included officers' patrols at night to enforce the curfew order, and a large number of guards and picquets. There was very little trouble in the city, however, and the inhabitants welcomed the new regime.

Major Ivens rejoined from hospital on March 30th, and on May 16th was relieved in command by the new commandant, Lieutenant-Colonel A. D. Cox. Leave was opened at the beginning of April to men and officers. During April and May opportunity was taken to reorganize and put in some much-needed training.

Advance Beyond Baghdad All this time General Maude had been pressing the Turks back north of Baghdad along both banks of the Tigris, and also up the Diala River. By the end of April the British advanced positions were at Samarrah and Shahroban, some seventy miles north and north-east of Baghdad respectively, and Baghdad was more or less safe from any danger of re-capture.

On June 18th the Regiment, under orders to Baqubah, handed over to the 110th Mahrattas, and crossing to the left bank, marched up the Diala under Captain Anderson, Colonel Cox being in hospital.

The stages were 19th Cassels' Post (13 miles), 20th Cunningham's Post (10 miles), and 21st Buhriz via Baqubah (18 or 19 miles). Owing to the heat by day early starts were made, and on the 19th and 20th the Regiment reached camp at 7.30 a.m. On these days no men fell out. The last day's march to Buhriz was more strenuous, and the camping ground at Buhriz was not reached till 9.30 a.m., after a hot and dusty march.

Baqubah Summer, 1917 At Buhriz, three miles from Baqubah, the Regiment was given an ample supply of European pattern tents, and had a comfortable camp close to the River Diala, which afforded some excellent and much-appreciated bathing.

From June to October the Regiment was employed in digging a defensive line of strong posts with communication trenches round Baqubah. They marched out daily at dawn to their work and returned to camp about 10 a.m. The temperature in July was very high, the highest recorded in camp being 127°, yet in spite of this the men worked remarkably well, and earned a great reputation in the Division as a good hard-working battalion.

They were also employed during part of this time on excavating a cutting for the projected railway, and completed this on September 19th.

The summer passed without any important incident. The remainder of the Brigade had joined the Regiment in Buhriz camp at the beginning of July. About this time the British forces were extending their influence eastwards from Baqubah, and an escort of 1 Indian officer, 30 rifles and a Lewis gun, which accompanied motor lorries to Beled Ruz, took part in a fight with Arab irregulars, but sustained no casualties. On July 29th a party of Arabs entered the camp transport lines and stole three rifles, and though the alarm was raised, they managed to escape in the darkness, firing a few shots with pistols at men who tried to intercept them. Hostile aircraft passed over the 36th Brigade camp several times during August, but did not drop any bombs on it.

At the end of August the effective strength of the Regiment was : 9 British officers (including the medical officer), 14 Indian officers, and 790 Indian other ranks.

In October the weather became consider-**Winter, 1917** ably cooler, and the British forces in Mesopotamia found themselves able to make various forward movements for the betterment of their positions. On the Diala line a column composed of the 36th Brigade (less the 26th Punjabis) left Baqubah on October 15th and marched to Shahroban. The 26th Punjabis took over garrison duties at Baqubah on both right and left banks, and consequently not only did they miss the operations for the capture of the Jabal Hamrin, but also were so busy that no training was possible.

This spell of duty lasted till December 12th. During this time life was rather monotonous. Occasionally a hostile aeroplane flew over camp, but did no damage. Two large drafts were received in October and November, and on November 3rd the Regiment changed camp to Shiftat, which was one and a half miles nearer to Baqubah than Buhriz. On November 26th one company was sent on detachment to Abu Jisrah.

An abundance of fresh fruit and vegetables was always obtainable locally from the luxuriant gardens at Baqubah, and was much appreciated by all ranks. Rounders and football were played, but lack of sticks and of a suitable ground did not admit of hockey. The British Officers were able to get excellent sandgrouse shooting during August and September, and later a certain number of black partridge and hare were also to be found. As a result, the health of the Battalion was quite good.

On December 12th orders were received to rejoin the 36th Brigade at Kurdarrah on the north side of the Jabal Hamrin. The Regiment was relieved of guards and duties on the 15th by the 102nd Grenadiers, and on the 16th, under the command of Colonel Cox it marched to Abu Jisrah, and on the 17th to Shahroban.

The 18th was spent at Shahroban, and on the 19th the march to Kurdarrah was resumed, crossing the Jabal Hamrin on the way. Everybody was delighted to see hills again after two years of nothing but the flat plains of Mesopotamia. The weather kept pleasantly cool, and on the 19th was very cloudy.

Immediately after arrival at Kurdarrah, whilst the transport was collected in the camp being unloaded, a hostile aeroplane suddenly appeared out of the clouds, flying very low (approximately 1,000 feet). The Regiment and its transport was all massed in a very small area, and must have offered a wonderfully good target, yet, strange to relate, it was neither bombed nor machine-gunned.

Mirjana On the 20th the Regiment marched to Mirjana through steady rain, and over a very bad and heavy road. On arrival there, wet and muddy, two companies took over the bridge-head defences on the right bank from the 62nd Punjabis. The two other companies and regimental headquarters went into camp on the left bank with the rest of the Brigade. Work was commenced at once on the defences, *i.e.*, lunettes and support posts on the right bank, and shelter trenches round camp.

On December 23rd the remainder of the Regiment crossed to the right bank, and on the 24th a half-company under Lieutenant Hunt was sent on detachment to Tel Burdan towards Qarah Tappeh as escort to a survey party. Apparently his escort duties were not very exacting, for he was able to send in a present of a goose to the mess on Christmas Day.

1918 During this period the Turkish forces were very careful to keep at arm's length. They left no garrisons or posts within reach which could be surrounded or destroyed by a sudden advance from our side. Consequently the opposing forces seldom came in contact. Hostile aeroplanes occasionally flew over, and sometimes dropped bombs, but generally these bombs did little damage. Our Light Armoured Motor Battery cars used to reconnoitre considerable distances, and on one occasion, on February 12th, one had to be abandoned on the road towards Qarah Tappeh. That night the 26th Punjabis and a squadron of cavalry were ordered out to cover its withdrawal, and it was successfully rescued.

On February 2nd the Regiment had been relieved by the 82nd Punjabis from the right bank bridge-head defences, and had withdrawn to the Brigade camp on the left bank. Here it spent February, March, and the first part of April. Life here passed very pleasantly and quietly. A good hockey ground was made, a garden started, and shooting (" black " and duck) was plentiful.

Persia. On April 8th orders were received to march to Qasr-i-Shirin (over the Persian **Dunsterforce** Frontier), and on the 10th all guards and duties were handed over to the Norfolks. On the 11th the Regiment marched at 6.30 a.m. to Khaniqin, a distance of 20 miles, and on the 12th, starting at 6 a.m. to Qasr-i-Shirin, a distance of 22 miles. On arrival here orders were received that the Regiment was to garrison certain detached posts at Sar-i-Pul, Pa-i-Taq, and Taq-i-Girrah. The garrisons for these posts marched from Qasr-i-Shirin the following day.

This extension of the line into Persia was due to the despatch into that country of " Dunsterforce," commonly known as the " Hush-Hush Crowd." This force was originally intended to form a nucleus of officers for organizing the Armenians into a force capable of defending their own country. Its despatch and maintenance in mid-winter over a country ill-found in communications proved so difficult that delays were unavoidable, and " Dunsterforce " never reached its original destination.

Its lines of communication up to the north of Persia constituted a problem that taxed the resources of the Mesopotamian Expeditionary Force to its utmost, and used up large portions of the IIIrd Corps. It was on this duty that the 26th Punjabis were employed for the remainder of the war.

Ivens' Column On April 21st orders were received for the Regiment, less two companies, to concentrate at Sar-i-Pul, and there, together with a section of 26th Jacob's Mountain Battery and one squadron 14th Hussars, to form a force called " Ivens' Column." (Colonel Cox had taken over temporary command of the 36th Brigade and handed the Regiment over to Major Ivens.) This force was to operate against hostile Kurdish tribesmen (chiefly Sinjabis) in conjunction with friendly tribesmen (chiefly Kalhurs and Kalkannis). On the 22nd the column concentrated at Sar-i-Pul, and on the 23rd marched 18 miles, crossing the Zohab River, and joined up with the " friendlies." There was slight rain on the night of the 22nd/23rd, and on the two following nights ; in consequence, the march to the foot of the pass over the Kuh-i-Harri Range was bad. The rôle of the column was chiefly to give moral backing to the " friendlies," and in accordance with that object the mountain guns, six Lewis guns of the Regiment, and the Hotchkiss guns of the Hussars were pushed up to the crest of the Kuh-i-Harri at dawn on the 25th, leaving the rest of the column bivouacked at the foot. The guns and machine guns opened fire at extreme ranges on the main Sinjabi position, which was on the farther side of the valley. A certain number of the enemy were concealed on the lower thickly wooded slopes, and from these there was fairly heavy firing for a few minutes.

The noise of our firing, however, and the sight of our shells bursting in the direction of the enemy,

FOUR SNAPSHOTS IN MESOPOTAMIA AND PERSIA, 1916-18.

1. Officers in Senna Camp, Tigris, 1916
3. Turkish Field Gun captured by 26th, at Shumran Bend, Feb., 1917
2. Imam al Mansur, near Kut, 1916
4. Hut building, Kirmanshah, 1918

was all that was required to hearten the " friendlies." These now dashed down the hill to attack the Sinjabis, who being proportionately disheartened by our backing of their foes, did not await the onslaught. They fled so precipitately that both pursuers and pursued were quickly out of sight of Ivens' Column, which having achieved its object, waited a couple of days at the foot of the Kuh-i-Harri, and then marched back to Zohab on the 28th, and to Qasr-i-Shirin on the 29th.

On the 14th May one complete company ("D" Company) left the Regiment to form part of a new Battalion, the 1/152nd Indian Infantry, forming at Amarah. Lieutenants C. E. O. Ansell, and J. H. Lyall ("Hindenburg") were transferred with this company.

L. of C. On May 31st the Regiment moved again, concentrating at Pa-i-Taq, but only for **Duties** a few hours, as in the evening one and a half companies marched up to Taq-i-Girrah *en route* to garrison posts on the Karind—Kirmanshah road. On the 7th Headquarters moved up the pass to Taq-i-Girrah, which is situated at the Mesopotamian end of the long Surkhadiza Pass. The Regiment was now split up into detachments, stretching from Pa-i-Taq to Chashmeh Safid, only one and a half companies being with Headquarters at Taq-i-Girrah.

On June 24th Headquarters and one company from Taq-i-Girrah, and one and a half companies from the posts concentrated in the Bowanij Valley (5 miles north of Karind) in support of His Britannic Majesty's Consul from Kirmanshah, who was endeavouring to obtain a definite, but peaceful, settlement with the Kurd tribes. This object was achieved, and on the 26th the troops returned to their respective posts.

On July 6th Headquarters moved to Sarmil, situated at the farther end of the Surkhadiza Pass, whence, on July 28th, 1 Indian officer and 100 men of the 46th Punjabis, and 35 men of the 87th Punjabis, who had been attached to the Regiment since the autumn of

1916, left to rejoin their own regiments. They had served the 26th truly and well, thoroughly identifying themselves with it and its interests. Everybody was very sorry to see them go, being particularly sorry to lose Subadar Firoz Khan of the 46th Punjabis. This depletion was shortly afterwards made good by the arrival of two drafts totalling 1 Indian officer and 127 men.

In spite of the fact that 100 men were employed daily on road-making, it was found possible to do a little much-needed training and musketry at Sarmil. This was very opportune, as the recently arrived drafts consisted largely of recruits, who after their long journey up from the base were found to be very greatly in need of individual training.

With the exception of the despatch of a column of 100 men under Lieutenant Teape into the Bowanij Valley on August 9th to demonstrate in support of a friendly Kurdish chieftain, and the control of Jelu refugees (Nestorian Christians fleeing from Turkish oppression in the vicinity of Lake Urumiah), there was little of importance to record during the remaining months of the Great War.

These Jelus were first met between Mahidasht and Kirmanshah, and Lieutenant Heathcote was detailed to escort them to Chasmeh Safid. The Regiment had a good deal to do with these refugees while at Kirmanshah, shepherding them past the town, policing their camps, supervising their sanitation [*sic !*], and doling out rations. These duties were not liked, as the Jelu was popular with no one.

Kirmanshah The Regiment had been ordered to Kirmanshah on August 13th, starting on the 20th and arriving on the 25th. Here it came under the command of the Persian line of communications (" Percoms "), and took over guards, duties, and detachments from the 1/2nd Gurkhas. On September 16th camp was moved from the high ground on the immediate outskirts of the town to the vicinity of the main bridge over the Karasu River about four miles out, and a hut-building scheme, in preparation for the winter, was undertaken forthwith. The work of building, improving, and repairing the huts was carried on well into December, and, though the roofs leaked very badly at first, these huts were really very comfortable, and a great improvement on tents.

In August, especially during the march to Kirmanshah from Sarmil, a large number of men went sick with " Bombay fever," that type of influenza which spread all through the force. In September and October there was also heavy sickness among both British and Indian ranks, principally on account of malaria and stomach trouble. The winter at the high altitude (5,000 feet) of Kirmanshah was very severe. There was almost continual rain from November to April, and in January snow fell. The cold was bitter. Duties in Kirmanshah were heavy, and the long four miles into the town was a field of mud.

The monotony of the routine was alleviated to some extent by games. Both hockey and football grounds were made, and a very fair hockey team was trained, which was sent down to Baghdad to play in the tournament. British officers had some sport also from a little indifferent duck and snipe shooting.

In the midst of this routine news was received of the armistice with Turkey (November 1st), and later of the armistice in France; but this made not the slightest difference in the " daily round and common task."

The Regiment continued the same duties **End of** at Kirmanshah, and it was not really till **War** nearly twelve months later that it felt the effect of the cessation of hostilities— namely, when it returned to India in October, 1919.

The remainder of its service overseas **1919** is soon told. It stayed at Kirmanshah throughout the winter till April. Some officers left on demobilization or for other reasons—

among others Captain D. S. Warren, on December 28th, to take up the appointment of Staff Captain, Persian lines of communication staff. He had ably carried out the duties of Quartermaster of the Regiment since August, 1916, and was a great loss.

On April 14th the 26th Punjabis vacated its lines on the left bank of the Karasu River, and went into camp on the right bank, leaving the lines empty for the 79th Carnatic Infantry, who arrived in relief on the following day. On the 17th duties were handed over, and on the 21st Headquarters and three companies commenced the march to Baghdad, the fourth company in the various posts from Kirmanshah to Hamadan following independently. The marches, though long, were on the whole good, the roads generally affording good going and the weather being not too hot.

Heavy rain was experienced at Sar-i-Pul and Qasr-i-Shirin, at both of which places a day's halt was made. Headquarters and three companies reached railhead at Qizil Robat on May 2nd, and after a wait of three days entrained in two trains for Advanced Base (Baghdad), and arrived there at daybreak on May 6th. It was at once ferried across to right bank, and next day took over duties from the 96th Infantry and moved into their camp. The 4th Company, which was three days behind, joined Headquarters on the 10th.

Baghdad Here, at Advanced Base, the Regiment was employed on garrison guards and duties till October. These duties were principally in connection with the very extensive Ordnance Demobilization Depot, which was situated there, and included small detachments sent out for short periods to guard dumps of Government property, and Turkish prisoner-of-war camps.

On July 19th the Regiment took part in a large ceremonial parade to celebrate the victory of the Allies. The parade took place on the outskirts of Baghdad city on the left bank, and the Regiment had to leave camp at 4 a.m. to reach the ground in time. Beyond this there was nothing of importance during the summer months. The weather was very hot during July and August, and averaged 115° by day. Luckily the nights were comparatively cool.

At last on October 2nd orders were received to be prepared to move to the base. All ranks were delighted at the prospect of getting back to India with the likelihood of an early return to their homes. On the 5th the first party, consisting of " B " and " C " Companies, under the command of Captain Jessop, departed down the Tigris on *S*.42, and were followed by the remainder of the Regiment on *P.S*.60 the day after. On the 12th both parties arrived at Basrah and went into camp at Magil. The 13th was spent in returning equipment, etc., and on the 14th the whole regiment embarked and sailed on H.T. *Bamora*, disembarking at Bombay at 9 a.m. on October 21st. In the evening of the same day the Regiment left in two trains for Bareilly.

This may be said to conclude the Regiment's active services during the Great War. It had truly borne much of the burden and heat of the Mesopotamian campaign. Though it was not there in time to share in the victorious advances of 1914 and 1915, it arrived at the beginning of 1916, when the fortunes of the British forces were at their lowest ebb, and took part in many of the actions which eventually led to the capture of Baghdad. The actions in which it was most closely engaged and suffered the heaviest casualties were the " Attack on the Dujailah Redoubt," March 8th, 1916 ; " Fighting in the Dahra Bend," January 26th, 1917 ; and " The Battle of Shumran Bend," February 24th, 1917. In all of these it acquitted itself right well, and earned an excellent name.

After the capture of Baghdad the 26th Punjabis did not have the good fortune to be included in any of the larger operations resulting in the gradual elimination of the Turks from Mesopotamia, but it carried out the equally burdensome and less glorious rôle of outpost and garrison duties for another two and a half years.

The Details of Casualties in Appendix VIII, and of Honours and Awards in Appendix VII, give together a measure of the Regiment's suffering and a gauge of its creditable performance.

CHAPTER IV

THE POST-WAR PERIOD, 1919-1923.

Bareilly   On arrival at Bareilly on October 23rd, 1919, the Regiment was amalgamated with the Depot, and immediately commenced demobilization.   About 380 Indian officers and men were rapidly demobilized.   At the same time the remainder of the Battalion was permitted to go on two months' war leave.   Thus, during November and December there was only a nucleus of the Regiment in Bareilly, and very little training was possible. In spite of the paucity of numbers, several entries were made for the Bareilly Brigade sports, and some successes were gained.   In a Brigade musketry competition, also, a team of four Indian officers secured second place, the " Queen's " taking first place.

Peshawar   Early in 1920 orders were received to be prepared to move to Peshawar, and the Regiment left Bareilly on February 10th, preceded three days earlier by the advance party under Captain H. L. C. Robertson.   It reached Peshawar on February 13th, and went into the Hari Singh lines, on the Jamrud Road.   Here the process of demobilization was continued.

No inspection of the Battalion had been carried out in 1919, but in 1920, in March, the General Officer Commanding the Brigade at Peshawar inspected it before it had any chance of training.   As was to be expected, he found that training was backward.

From February to December, 1920, the only training attempted was individual training with some training of sections and platoons.   It was intended to go into camp later on for company and battalion training ; but, just before that training commenced, orders were received to " stand by " for Waziristan.

Though Peshawar at this time was not a very peaceful station, it was with some consternation that we heard we were again proceeding on field service. All ranks were rather tired of the war and wanted a rest. In Peshawar duties were very heavy and night raids numerous, so that it was not altogether restful; but men and officers had comfortable lines, and leave, though restricted, was open.

On one occasion in Peshawar a guard furnished by the 26th Punjabis on the Station Hospital, and commanded by Naik Karm Ilahi, acquitted itself well. On being fired upon by thieves, it sallied forth, took up a position, and succeeded in wounding one of the attackers.

While in Peshawar two distinguished Indian officers of the Battalion were highly honoured by H.M. the King-Emperor, in being promoted to the honorary rank of Lieutenant. The first was Subadar-Major Ishar Singh, who had again returned to his place as Subadar-Major, and had earned the Order of British India, 2nd Class. The second was Subadar Jan Gul, who through his gallant and loyal behaviour with the Afridis in East Africa, had won the Indian Order of Merit. Both these officers retired shortly afterwards on pension, to the great regret of all ranks, British and Indian.

Another loss to the Battalion in Peshawar was the death in hospital there of Captain C. C. T. Teape, who had only recently rejoined from sick leave in the United Kingdom.

Waziristan Definite orders were received on December 23rd for the 26th Punjabis to leave Peshawar for Waziristan about the end of January, and on January 16th telegraphic orders came to entrain for Mari-Indus on the 20th *idem*.

The Battalion, with a strength of 9 British officers, 18 Indian officers, 591 Indian other ranks, and 55 followers, under the command of Major E. A. Maude, D.S.O., reached Mari-Indus on January 21st, and on

23rd and 24th proceeded by train to Tank, whence it marched to Kotkai, arriving there on January 26th. Here it took over permanent picquets and road-protection duties from the 2/90th Punjabis, and immediately took its part in the usual incidents pertaining to frontier warfare.

On January 29th the Battalion, leaving camp at 5 a.m., co-operated in the establishment of New Pioneer Picquet. That night Gujar Picquet was heavily sniped, and on January 31st the watering party to Chalk Hill Picquet was sniped, both without casualties.

On February 6th Lieutenant-Colonel A. D. Cox arrived and took over command. During this month duties and fatigues in camp were heavy. Much discomfort was also experienced from the prevalence of wind, and on February 28th some of the tents were actually blown down by the force of the wind in Kotkai Camp.

On March 15th 1 man was wounded at Spinkai Picquet by a sniper, and on the 17th a group of 1 non-commissioned officer and 6 sepoys, moving out from camp to take up a day position, were heavily fired on from close quarters, and lost 5 men killed and 1 dangerously wounded.

On March 19th road-protection troops engaged about eight armed Mahsuds, and claimed to have hit two. Another party clearing sangars also fired on some Mahsuds, one of whom was reported wounded. On March 29th and 31st, at New Thorny Ridge Picquet and Nai Kach Picquet, there were slight affairs in which we had a Lewis gun struck by a bullet and 1 Mahsud was wounded.

On April 10th, at 2 a.m., " A " Camp Picquet was heavily sniped from close quarters, but dispersed the enemy with rifle grenades and rifle fire.

The above few incidents were typical of the ordinary routine existence of a battalion protecting part of the line of communications in the Takki Zam. Active operations were nominally over, and the line up to Ladha was effectively established. Convoys moved up and down daily. The system of protection was partly by permanent picquets on the main commanding heights, but this was supplemented by " road-protection " troops (about two companies) which moved out daily before the convoy started, and took up subsidiary positions overlooking nullahs and minor features.

These duties were fairly arduous, especially in the winter, when men had to wade the Takki Zam in the cold early morning, and then sit all day in their wet clothes. Occasionally with the Takki Zam in spate, the wading was not only cold, but dangerous.

Sorarogha In April orders were received for the 26th Punjabis to move to Sorarogha in relief of the 1/69th, being themselves relieved at Kotkai by the 1/29th Baluchis; and the Battalion marched to Sorarogha on May 1st. Major E. A. Maude took over command again from Colonel A. D. Cox, who proceeded on leave pending retirement.

Colonel Cox had served originally in the 69th Punjabis and latterly in the 21st Punjabis. He joined the Battalion as Commandant at Baghdad in 1917, and commanded it for most of the remainder of the war. For a short time he acted as Brigade Commander in Persia. He maintained the efficiency of the Battalion at a high standard, and for his services was mentioned in despatches and promoted full Colonel.

About this time the composition of the Battalion was changed. A company of Dogras from the 37th Dogras and a company of Jats from the 47th Sikhs replaced our Afridis, some Sikhs, and a company of Punjabi Muhammadans. The Dogras were a fine body of men, well trained and efficient. The Jat company contained good men also; but had suffered in training by having been passed on from one battalion to another. Dogras were not new to the 26th, as there had been a company in the Regiment twenty years previously. Jats were new, but were well known from their record in the Great War. While regretting the final loss of the Afridis, who had been a feature of the Regiment for so long, there can be no doubt that the latest composition of the Battalion is hard to beat.

At Sorarogha the Battalion took over its share of periodical tours of duty as follows :—

1st Period (16 *days*).—2 companies down-stream picquets ; 2 companies road-protection (3 days out of 4).

2nd Period (8 *days*).—Mobile column, ready to move at half-hour's notice. " Training," sometimes called " Rest." Up-stream road-protection ; 2 days, 2 companies each day.

3rd Period (16 *days*).—2 companies up-stream picquets ; 2 companies, alternate days on Brigade guards and duties.

4th Period (8 *days*).—Same as 2nd Period.

During May, June, and July Sorarogha Camp was sniped several times at night. During these months also much discomfort was experienced from strong wind and dust, and from heavy thunderstorms, which caused the Takki Zam to come down in spate.

The next three months constituted the unhealthy period in Waziristan, and the sick rate rose to very high figures on account of malaria. At one time it reached a daily sick rate of 200, with about 50 admitted to hospital. This threw very heavy duties on the remainder. The Jats particularly found the climate trying, partly owing to their inability to obtain milk and vegetables. Many of them, while debilitated by malarial fever, succumbed to pneumonia as the cold weather came on.

In December the 26th Punjabis were detailed as reserve to the Wana Column, and stood ready to move as strong as possible at twelve hours' notice. Occasional sniping of camp at night continued, and sometimes the parade ground *raghza* (plain) was sniped of an afternoon. The days chosen for this were usually those of some tournament when the whole camp had turned out to watch. On such occasions both players and spectators stampeded hurriedly for camp. On December 16th a watering party of the Jat Company proceeding to Bluff Picquet was ambushed *en route* and though they stoutly tried to drive off the Mahsuds with pick-helves (their only weapons), the enemy managed to make off with the mules after wounding the lance-naik in command, who was later awarded the Indian Distinguished Service Medal. Bluff Picquet opened fire, and one mule was killed and another returned to camp. The next day the up-stream convoy was sniped, but the enemy was driven off by fire from our picquets.

Thus passed 1921. The men, in spite of the hardships, were fairly content. They were getting free clothing, batta, and a field service scale of rations ; and leave, though limited, was not closed. They were living in E.P. tents, and had as much kit as they desired. The climate, with the exception of the malarial season, was healthy. Vegetables, milk, and luxuries were sometimes obtainable, but not abundant. The officers' lot was, in comparison, not so pleasant. There was no sport, as shooting was forbidden ; and though games were easily to be had and there was an apology of a club, there was nothing like extra pay or batta to compensate them for extra expenses. Free rations hardly counted, for it would have been quite impossible to obtain supplies locally in a barren country such as Waziristan.

1922
Minor
Operations

The year 1922 opened with a couple of unfortunate incidents. On January 12th, in a thick early morning haze such as often arose, some Mahsuds ambushed a party of our road-protection troops (Dogra Company) who were moving across the bush-covered *raghza* to take up No. 3 Picquet position. The Mahsuds were cleverly concealed and suddenly opened a large volume of fire.

Five of our men were killed outright, 6 died of wounds, and 2 more were wounded. Nine rifles and sets of equipment were carried off by the Mahsud knifing party, covered by other firing parties. Marble Arch Picquet and No. 2 Day Group opened fire on the enemy, and inflicted at least one casualty; but the enemy's line of retirement was open, and they soon melted away into the distance.

For this mishap no heavy blame could be attributed to anyone. It was due to a combination of unfavourable circumstances. The only lessons to be learnt were greater caution, and closer support from a second line. In this case the second line was also engaged by the enemy; but the bush and fog rendered visibility very bad.

On January 16th the day-group going out from Flat-head Picquet was ambushed in a fold of ground, but, owing to prompt action on the part of a sepoy, who was in command, though 2 men were killed, no rifles were lost. This man, a Jat, received the Indian Distinguished Service Medal. These two incidents made all ranks extremely cautious.

On January 20th Bushy Hill Picquet, acting on information telephoned from camp, opened fire on a party of Mahsuds and caused two casualties. Two new picquets, Dean and Ellis, were established this month on the line of the new road that was to be constructed for motors; and the Battalion took part in covering their establishment. This new road was the chief reason for the maintenance of the Waziristan force on the Takki Zam line, so it was a matter of great interest.

The winter at Sorarogha was very severe. The normal temperature at night was well below freezing point, and, in addition, there were many rain-storms and several days of continuous rain. On January 27th the up-stream " road-protection " troops went out in rain and sleet, and what with their feet soaked by repeated wading through the cold water of the Takki

Zam and the drenching from the rain, they spent an uncomfortable day picqueting the road.

On February 8th Lieutenant-Colonel P. S. Stoney took over permanent command of the Battalion from Acting Lieutenant-Colonel E. A. Maude, D.S.O., who proceeded home on leave.

The spring of 1922 saw several changes **Organization** which affected the Regiment closely. The first was the reorganization of the Indian Army into groups of four or five field battalions with a training or depot battalion. In accordance with this scheme, our depot, which had been at Jullundur, was moved in April to Sialkot, and amalgamated from May 1st with the 29th Punjabis, which became the training battalion (T.B.) of the group. The group, now called " Regiment," consisted of the 25th, 26th, 27th, 28th, and 29th Punjabis, known henceforward as the 1st Bn. 15th Punjab Regiment, 2/15th, 3/15th, 4/15th, and 10/15th respectively. This scheme, though depriving the battalions of the power of selecting and training their own recruits in peace, allowed for greater elasticity in the provision of recruits and expansion of units in war, and also obviated the hurried reorganization of a unit, which the formation of a depot on mobilization always entailed.

The second change was the reduction in the number of British officers : this was necessitated by units being disbanded and the authorized strength of the remaining units being decreased. During the course of a few months almost the whole body of British officers was changed. At this time the strength of British officers in the 26th Punjabis was thirty-three (*vide* Army List of April, 1922), and six more were posted in the next few months from the disbanded 2/66th Punjabis, which Lieutenant-Colonel Stoney had raised and commanded. It was decreed that, of officers commissioned during the war, about 250 had to be demobilized, and this meant that about every other one must go. Several officers desired to go on the terms offered, but others had to be retired compulsorily.

Even after this reduction the Battalion was still over strength in officers. The authorized strength was only twelve. In order to try to retain officers on the rolls, many were recommended for staff appointments, which they were well qualified to fill. At the end of 1922 the Battalion had an officer in a billet in almost every post up and down the Takki Zam line in Waziristan. These billets included :—G.S.O.3 Wazir Force ; Field Cashier, Sorarogha ; S.S.O. Kotkai ; R.T.O. Khirgi ; Divisional Signals, Dera Ismail Khan (2) ; and M.F.O. Sorarogha. We also had Major Ivens, Recruiting at Jullundur ; Major Anderson, with Indian State Forces ; and Captain Farwell, Adjutant of a Railway Battalion.

In the meantime the Battalion at Sorarogha carried on its routine duties, which were by no means devoid of incident.

On March 8th, while two companies of the Battalion were on down-stream road-protection, and the other two in picquets down-stream, enemy snipers from near " Marble Arch " held up the convoy and killed two camels. By next morning these two dead camels had been completely eaten up by hungry Mahsud villagers who lived in caves near where the camels fell ! Marble Arch Picquet was also molested at a later date by Mahsuds sniping the watering parties : two mule saddles were hit. In retaliation the picquet fired on and wounded a Mashud, who, they said, behaved in a suspicious manner one evening not far from the picquet. As usual, the Mahsud claimed compensation, and asserted that he had no evil intentions. The Company Commander, Captain Steer, who had encouraged the retaliation, narrowly escaped having to pay this compensation.

At the beginning of June the new motor road over the *raghza* past Dean and Ellis Picquets was opened, and the old road in the river bed closed. New picquets had been established on " Dazzle," " Duke's Nose," and " Plateau," and the old picquets at " Bushy Hill," " Marble Arch," and " Flat-head " were withdrawn. Several cases of sniping occurred in the vicinity of " Dazzle," but we suffered no casualties.

On July 1st, the head of the up-stream **Ahmadwam** convoy was ambushed, near Ahmadwam, **Ambush** and thirty mules carried off. Havildar Kaku, " D " Company, was killed. As it was *Khasadar* (irregulars) day, none of the men of the watering parties were armed, so they were unable to offer any resistance. The *Khasadar* protection proved useless, and it was only fire from our own permanent picquets which prevented more of the convoy being looted. Three permanent picquets were in sight of, and about 1,000 yards distant from, the locality of the ambush. Each of these picquets, simultaneously with the attack, were fired upon from an opposite direction by enemy covering parties. This showed how carefully the attack had been planned. In the convoy, besides the loss of mules, Captain Edmiston, transport officer, was mortally wounded.

As a reprisal for this attack, the village of Ahmadwam, overlooking the site of the ambush, was surrounded, on the night of July 6th/7th, by a concerted movement carried out by units of the 21st Brigade from Sorarogha, and all male inhabitants were brought captive to camp. The 26th Punjabis, with Headquarters Company and " B " Company (the remainder of the Battalion being on duty), carried out the task of seizing the village and capturing the men, while the 1/4th Gurkhas and the 106th Pack Battery occupied the surrounding heights to prevent outside interference. The prisoners, twenty-seven in number, were detained for two months, and the village fined Rs3,000 for not giving warning of the attack. No actual complicity could be proved to justify any death sentences.

On July 15th raiders attacked some local guards just outside " C " Company's lines at Sorarogha, and

FOUR SNAPSHOTS IN WAZIRISTAN, 1922-23.

1. Some British Officers near Sorarogha, 1922. Lieut.-Col. Stoney, standing; seated left to right—Lieut. Bwye, Capt. Fulton, Capt. Robertson (Adjt.), Maj. Edwards (2nd in Command)
2. A Vickers gun nest at Haidari Kach
3. Dargai Obah, Camp of 9th Brigade Column, 1923
4. Col.-Comdt. Keily addressing Indian Officers on departure of battalion from Wazirforce, April, 1923

carried off about 700 sheep belonging to a contractor; several of the shots fired came over camp; but there were no casualties in camp.

An example of the difficulties with which one has to contend in a perimeter camp was afforded by an incident which occurred one day this summer. A portion of Sorarogha camp, given over mostly to incinerators, was precipitous on its outer sides, and was therefore only lightly protected at night by a moving patrol. At dusk it was noticed by some observant individual that an incinerator chimney standing close to the edge of the cliff had a rope attached. On investigation it was found that a rope ladder was hanging down the precipice. An ambush was accordingly arranged in order to catch any thieves coming in or going out of the camp; but, unfortunately, the thieves must have realized that their contrivance had been detected. They tried again another night, but the vigilance of our patrols again prevented them.

In August and September there were heavy storms and resultant spates. Sickness again increased, but though it averaged about fifty daily, with about forty in field hospital, it never rose to the same figure as in the year before. This was partly due to a better supply of vegetables. The S. and T. supply was augmented from a large regimental garden, which flourished exceedingly. At the end of October malaria began to decline again.

Winter Operations About this time preparations commenced for a withdrawal of regular troops from the line above Jandola, and for the simultaneous establishment of a large camp at Razmak, near Makin. All these preparations were kept very secret; but, as officers had to reconnoitre very carefully the areas allotted to their units for covering the withdrawal, and the men had to reduce their kit to Field Service scale, there were not many persons in the force who did not know about the withdrawal. The actual date thereof was more carefully guarded ; but even the Mahsuds were able to make a very shrewd guess at that ; and it was only due to constant change of plan, caused apparently by indecision among the " powers that be," that the Mahsuds' forecast was not exactly correct.

In the end, after back-loading tons and tons of tents, stores, and kits, the whole plan was altered and the line was maintained by regulars for a further period of nine months, while the motor road was continued through Razmak to the Bannu line at Tal (near Idak).

The 26th Punjabis were saved the disappointment of having to stay in Waziristan for this extra period, by being one of the first units due for relief. A few months' delay was more easily borne then the prospect of another summer and autumn in Waziristan.

In the meantime the situation among the Mahsuds was distinctly deteriorating. According to some opinion, the vacillating policy was responsible, while other opinion held that the bombing of villages exacerbated the doubtful tribes. Eventually it was decided to punish Makin, and also to withdraw the South Waziristan Scouts from Wana. Incidentally terms had to be arranged with the tribes for providing labour to construct the motor road through Razmak and through the Shahur Tangi.

1923 The 26th Punjabis took no part in the operations against Makin, except in maintaining the line of communications near Sorarogha. By January 1st, 1923, the Battalion was on Field Service scale, ready for operations. Owing to the departure of other troops from Sorarogha for Makin, the duties at Sorarogha became heavier. Several cases of sniping occurred at road-protection troops, at picquets, and at troops exercising. In one case we had one man killed (a Dogra), and in another one man wounded.

The Makin area was heavily bombed by aeroplanes, and by howitzers situated in Ladha camp, continuously day and night from January 7th to 12th. in the hope that this would force the inhabitants to give up Musa Khan, the leading irreconcilable. This failed, however ; and on February 1st the Ladha Brigade moved up to Makin and commenced devastating the area.

One of the most trying duties for the Regiment at this time was road protection up-stream. The river swelled to an unusual depth in January, and remained swollen for months. This interfered with the convoy of supplies, especially of sheep, and entailed much labour on the upkeep of the road. The Pioneers, busy elsewhere, were unable to divert the water off the road, so the 26th Punjabis volunteered to try. They made many valiant efforts, which were often successful for a time, but their efforts were eventually always marred by greater floods. During this period several men were nearly drowned, and three rifles were lost in the floods.

On one occasion two men were crossing a swollen stream and were holding each other's hands for greater stability. When nearly across, both stumbled and were being rolled over by the force of the stream, when Captain Conner pluckily dashed in and pulled them out. The fact that Captain Conner was himself unable to swim rendered the act all the more commendable.

At the end of March a brigade from the Bannu side, having been firmly established at Razmak, and the tribesmen having agreed to the construction of a road through Makin, the 9th Brigade (Ladha Brigade) withdrew down to Jandola and prepared to march up through the Shahur Tangi to Wana.

The 26th Punjabis were detailed to join this column at Jandola, and on March 30th moved there direct (15 miles) in one day without halting at Kotkai. They had handed over their duties at Sorarogha to the 69th Punjabis.

On April 1st the Wana Column, under the **Wana** command of Colonel-Commandant Keily, C.B., **Column** D.S.O., started for Wana to relieve the Scouts there. On the first march to Chagmalai the

26th Punjabis were part of the Advanced Guard, and were detailed to seize the south side of the Shahur Tangi. A few shots were fired by snipers, but otherwise no resistance was offered.

An amusing incident occurred that morning. The Dogra Company brought in to the Commanding Officer a Gurkha and a Mahsud, the latter heavily guarded by our men with fixed bayonets. The Gurkha, who was an ex-rifleman and now a private servant, explained that he had gone out to fight the enemy and found this Mahsud coming to attack the army, and had bravely captured him. The Gurkha was about four foot nothing, and had no other arms but a fly-whisk, while the Mahsud was about six foot and had a dangerous-looking felling axe. The situation seemed difficult to understand, but was explained eventually by the fact that the Gurkha was mad and the Mahsud quite friendly !

Some permanent picquets were established that day at the eastern entrance to the Shahur Tangi, and next day others were established as far as the western exit. The third day the column marched through to Haidari Kach, leaving a garrison at Chagmalai.

Haidari Kach
The 26th Punjabis were left at Haidari Kach on the 5th, when the remainder of the column marched on to Sarwakai. Then the whole column halted for four days while all the baggage camels and supply column camels were sent back to Jandola to fetch up more supplies. In the meantime the battalions at Chagmalai and Haidari Kach held the permanent picquets and kept the road open.

While at Haidari Kach the Battalion was assisted in its duties by a body of South Waziristan Scouts at Dwatoi, and by a collection of *Khasadars* who camped in the old fort on the left bank opposite our camp. These *Khasadars* were sometimes more of an anxiety than a help. They had no uniform, and the only way they could be recognized was by their possession of white flags presented by the political agent. Even the addition to the flags of a distinguishing mark sewn on by our regimental *durzi* did not quite preclude the possibility of these flags being misused.

These *Khasadars* were sometimes used to picquet the farther hills beyond those on which our own picquets were placed, or were sent on ahead of our picqueting troops to see if the way were clear. On the whole, they worked fairly well, and saved a certain amount of extra picqueting ; but they were never very dependable.

On one occasion a party of half a dozen *Khasadars* sent to occupy a hill returned to report that after a severe struggle they had been driven away by the enemy. No shooting had been heard, and on further inquiry it was ascertained that the struggle had only been one of words ! Fortunately the enemy took no further action against us.

Completion of Operations

On April 10th the 26th Punjabis, being joined by the 2/50th Kumaonis from Chagmalia, and having handed over the permanent picquets to the South Waziristan Scouts, marched on to Sarwakai to join the column. The column, with its supply column refilled, then marched on through Dargai Obah to Rogha Kot. The march to the former place was extremely difficult for camels, as the rain poured down all day and made the descent into camp down a steep hill path very slippery and precarious. From Rogha Kot the column moved up towards Wana and picqueted the steep sides of the Bobal Mountain and Ghiza Pezha ridge in order to protect the South Waziristan Scouts, who marched out of Wana to join the column.

On April 14th the whole column withdrew to Dargai Obah, and on the 15th to Sarwakai. Here the South Waziristan Scouts, assisted by all available troops, had built themselves a strong fort and were firmly established. Leaving them there, the column then marched back to Haidari Kach and thence to Jandola.

All through these operations battalions **Lessons** were detailed to open a certain section of the road by picqueting both sides. The next battalion then moved through and opened the section beyond. As soon as all the baggage was through, the first battalion withdrew its own picquets and then, handing over rear-guard duties to the battalion next beyond it, marched on to camp. Each of the other battalions then withdrew in turn. After passing Sarwakai no permanent picquets, except camp picquets, were established. The first battalion frequently moved off before daylight, but otherwise there was not any night-marching.

For the establishment of permanent picquets elaborate arrangements were usually made. In addition to the personnel of the picquet, which varied in strength from a couple of sections (say 12 men) to a platoon (say 25 men), separate parties were detailed for each of the following duties :—Building (usually Pioneers if available), carrying stones, erecting wire entanglement (Sappers and Miners), and for protecting the working parties. On one day's march the 26th Punjabis were detailed to build and occupy six permanent picquets. Though another battalion, acting as advanced guard, had opened the road, local protection had to be afforded, and in the absence of any Pioneers building had also to be done. This severely taxed the strength of the unit, and except for an attenuated Battalion Headquarters, and carefully-reduced baggage guards, it left no spare officers or men down below. Usually one company could not manage the covering and construction of more than one permanent picquet. Special stores for picquets, such as wire, picquets, water tanks, signalling equipment, ammunition, Véry lights, were carried under Brigade arrangements and distributed as required.

From beginning to end there was hardly any serious opposition. There was occasional sniping, and when-

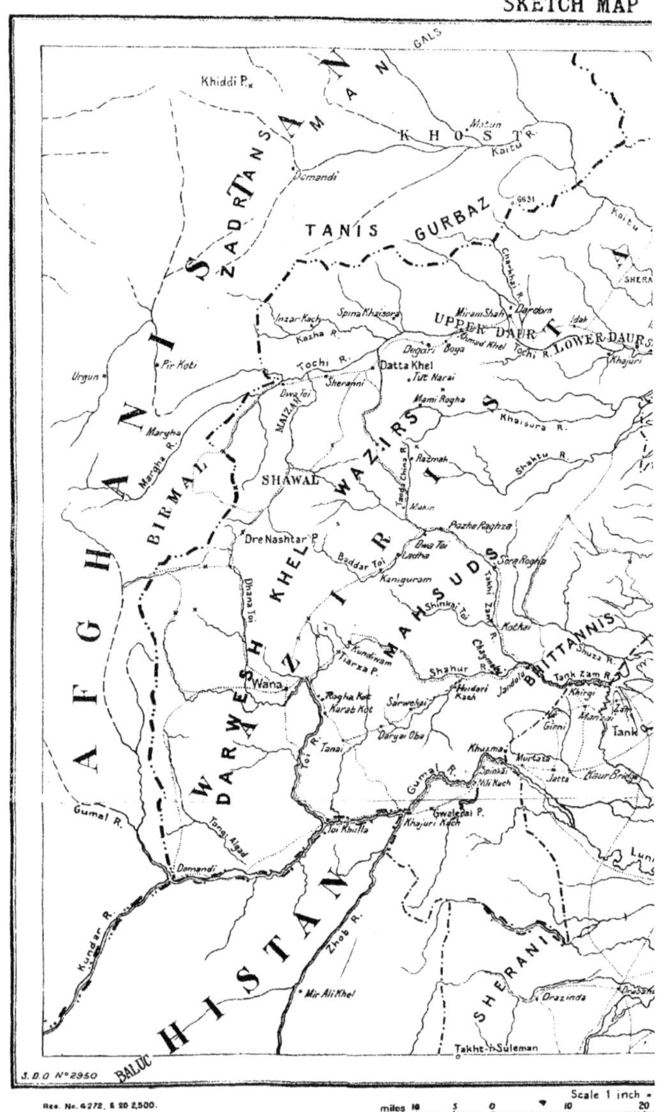
GALS
Khiddi P.
KHOST
Motun
Kaitu
ZADRAN
SAMAN
Domandi
TANIS
GURBAZ
6631
Kaitu
Charkhai R.
SHERA
Inzar Kach
Spina Khaisora
UPPER DAUR
Miram Shah
Darobm
Idah
Kazha R.
Ahmad Khel
Tochi R.
LOWER DAURS
Degari
Boya
Pir Koti
Tochi R.
Datta Khel
Tut Narai
Khajuri
Urgun
Sheranni
Mami Rogha
Dwa Toi
MAIZAR
Khaisura R.
Margha
WAZIRS
Taogi China R.
Razmak
Shaktu R.
Margha R.
BIRMAL
SHAWAL
Makin
Pazhe Raghza
Dre Nashtar P.
Dwa Toi
Ladha
Sora Rogha
Baddar Toi
Dhana Toi
KHEL
Kaniguram
MAHSUDS
Takhi Zam R.
Shinkai Toi
Kotkai
AFGHANISTAN
Kundiwam
Tiarza P.
Shahur
Chagi R.
BHITTANNIS
Shuza R.
Tank Zam R.
Wana
Shahur R.
Haidari Kach
Jandola
Khirgi
Rogha Kot
Sarwekai
Zam
Karab Kot
Manzai
DARWESH
Daryai Oba
Girni
Tank
Toi R.
Khuzmai
Murtaza
Tanai
Gumal R.
Spinki
Mili Kach
Jatta
Kaur Bridge
Gumal R.
Gwalerai P.
Khajuri Kach
Toi R.
Luni
Tangi Algad
Loi Khulla
Domandi
Kundar R.
Zhob R.
SHERANI
Orazinda
Dre Sh
BALUCHISTAN
Mir Ali Khel
Takht-i-Suleman
S. D. O N°2950

Scale 1 inch =
miles 16   5   0   10   20

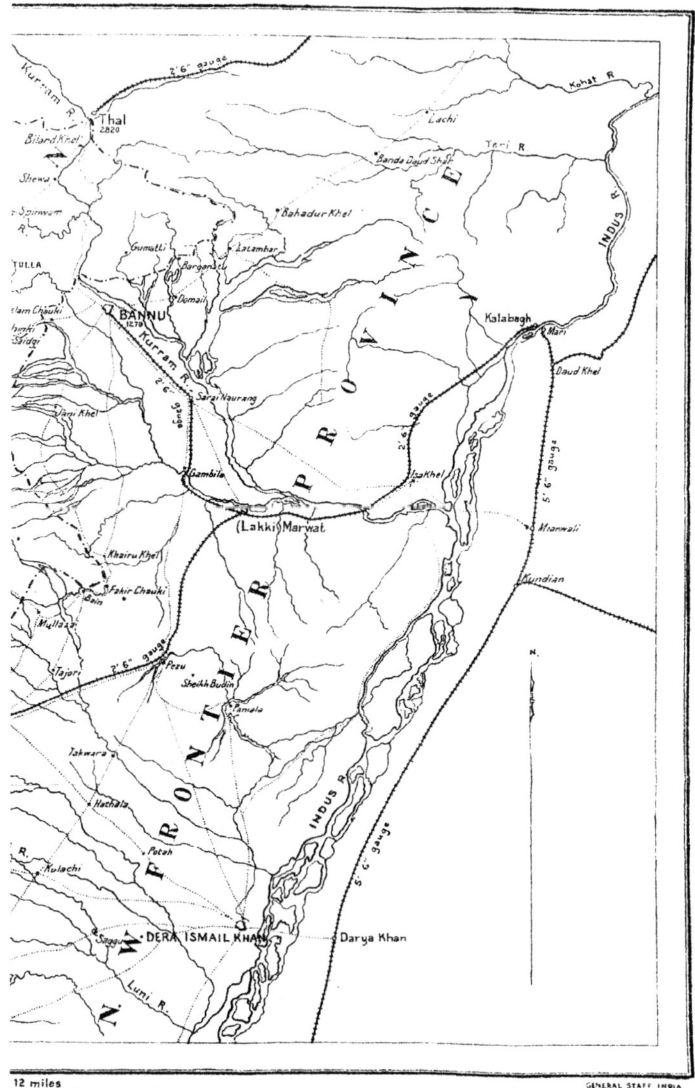

12 miles
30   40   50 miles

ever this occurred a large volume of fire was directed on the sniper. Our machine guns, under Lieutenant Bwye seized several opportunities of retaliation, but otherwise our men did not fire much. The Battalion was fortunate enough to escape any casualties ; in fact, they were seldom fired at ; perhaps, by wiser precautions, better concealment, and more rapid movement, they gave fewer opportunities to snipers than other battalions. Battalion Headquarters twice came under fire, but in each case it was while " liaisoning " with Brigade Headquarters.

There were also no casualties in the Battalion from admissions to hospital during the whole twenty days. This was a most creditable performance. Other battalions had many admissions. Possibly the fitness of the men was due to the keenness with which they had recently contested a platoon football tournament, and to the general hard work in which they had been engaged for the last three months.

Even the followers of the Battalion bore the hardships wonderfully well. They carried out the hot and trying marches in a formed body headed by the grey-bearded mess-bearer, Allah Jowaya, who strode along with a large stick at the head of what he called his " Platoon." He was the only man in the Battalion who could boast the Egyptian medals for the 1896 Suakin campaign ; but in spite of his age, he never spared himself on arrival in camp, and forthwith arranged for, and himself attended to, the comfort of the officers.

On the conclusion of the operations, and
Leaving before the Battalion left Jandola for Tank,
Waziristan the Colonel Commandant of the column collected all the Indian officers and British officers and bade them farewell. He said he knew that the 26th would acquit themselves well, and he had not been disappointed. He thanked Lieutenant-Colonel Stoney and all ranks for their loyal and willing assistance.

At Tank the 26th Punjabis took over duties from the 120th Rajputana Rifles for a few days, and on relief by the 6th Jats left Waziristan for Jhelum on May 2nd, arriving there on May 5th. Excepting for a short stay at Agra on return from China in 1915, and a short stay at Bareilly on return from Mesopotamia in 1919, this was the first time for over a quarter of a century that the Battalion had been quartered in a cis-Indus cantonment.

Thus ended two years of Field Service in Waziristan.

Both Colonel Dundas, commanding the 21st Brigade, and Major-General Sir Torquhil Matheson,* commanding the Waziristan Force, wrote letters expressing their regrets at the Battalion leaving their command, and thanking it for the " loyal and cheerful way in which all ranks had carried out their duties." They also recorded their " appreciation of the excellent work of the 26th Punjabis and of the keen and soldierly spirit invariably shown by the Battalion."

The excellent signalling of the Battalion was specially commended. With the numerous permanent picquets and the extended nature of all operations in Waziristan efficient signalling was all-important. Though there was no British officer to place in charge of the signallers after Lieutenant Lane left, there was fortunately a very good Indian officer instructor in the person of Subadar-Major Buta Singh. The latter took infinite pains to train them up to the high standard they had attained when he was a havildar and assistant instructor of signalling.

Moral A word about the moral of the Regiment will not be out of place here. The years since the war had been years of political unrest in India. Sedition was rampant in many cities and towns. It was almost impossible to prevent this insidious disease from spreading into the ranks

* For details of General Matheson's letter, see end of Appendix II.

Photo by

F. Bremner, Simla

BRITISH AND INDIAN OFFICERS, 2/15th PUNJAB REGIMENT, JHELUM, 1923.

Back Row, standing : Jemr. Bhalu, Jemr. Nazir Ali Shah, Jemr. Sharm Singh, Subr. Dasaundha Singh, Jemr. Bachan Singh, Subr. Gurditta, Subr. Sardar Khan, Jemr. Sher Dil, Subr. Narayan Singh

Second Row, standing : Capt. A. H. Pollock, M.C., Lieut. W. J. Blake, Lieut. B. J. Fitzpatrick, Capt. A. G. I. A. Goddard, Lieut. M. Bwye, Lieut. A. H. Rootes, Capt. T. S. Conner, Lieut. J. R. Thompson

Third Row, sitting : Capt. H. L. C. Robertson, Subr. Udmi Ram, Maj. E. Edwards, Subr. Maj. Buta Singh, Maj. E. A. Maude, D.S.O., Subr. Suraj Din, Maj. O. D. Bennett, Subr. Duni Chand, Capt R. D. Fulton, M.C.

In Front : Jemr. Lehru, Jemr. Kirpa Singh, Jemr. Motla

of units. The year 1922 was a particularly trying one for all loyal Sikhs.

Yet, in spite of it all, the 26th Punjabis managed to keep a healthy mind as well as the healthy body already referred to above. This can only be attributed to the excellent wisdom and example of the Indian officers, led by the Subadar-Major. It is far easier to foster sedition than to combat it, and the Indian officers accomplished the more difficult task. Ignoring the widely published incitements to non-co-operation, they co-operated most loyally with the officers and the Government whose salt they ate. They realized, as Indian officers often realize better than anyone else, that the Sarkar, whatever errors of judgment it may make, yet is always sincerely striving to do what it considers best for the country. Indian officers are in closer touch with Englishmen than their civil compatriots, and have learnt that they are held in high esteem and treated with all respect by their brother officers. This spirit of comradeship has been cemented by years of fighting side by side, and will continue to be fostered by mutual co-operation in training and games.

APPENDICES

APPENDIX I.

Alphabetical List of Officers of the Regiment.

NAME AND RANK	PERIOD SERVED	APPOINTMENTS, ETC.
Almond, H. R., 2/Lieut. (I.A.R.O.)	18/11/18 to 1919	
Amies, B. J., 2/Lieut.	19/11/15 to 1916	Attached.
Anderson, C. C., Lt.-Col. (S.C.)	23/10/84 to 1901	Wing Officer; Burmah Police, 1888; Off.-Comdt. Burmah Police Bn., 1889 ; Comdg. Depot, Jhelum, 1895 and 1900. Posted to 33rd B.I., 1901.
Anderson, G. W., Major	18/10/05 to date	D.C.O. ; Adjutant 20/11/11 ; with 128th Pioneers, Egypt, 1915 ; Actg. D.C.C., Mesopotamia, 1916 to 1918 (2 mentions) ; with 2/19th P. Waziristan, 1920, and with 26th, 1921 (mention) ; Indian State Forces, 1922.
Ansell, C. E. O., Lieut. (66th Punjabis)	Dec. 1917 to May, 1918	With Bn. in Mespotamia.
Atkins, W., Lieut. (Genl. List Infy.)	18/2/63 to 1863	
Badgeley, W. F., Lieut. (20th N.I.)	28/12/59 to June, 1865	Adjutant, 14/7/60 ; permitted to study at the Thomason College till 1/11/64.
Bagnal, R. L., Major (74th Punjabis)	23/1/21 to 9/3/21	Second-in-Command.
Bailey, C., Lt.-Col.	1886 to 1887	Attached for Burmah as Lieutenant ; served later in the 16th B.L., and was S. Captain at D.I. Khan, 1904.
Baird, E. H., Capt. (R.A.M.C.)	Aug., 1918	With Bn. in Mesopotamia.
Balkrishna, — (I.M.S.)	Aug., 1916 to Sept., 1917	With Bn. in Mesopotamia.

Name	Dates	Remarks
Bateman, G. C., 2/Lieut. (I.A.R.O.)	4/11/17 to 1918	
Barnes, J. A., Capt. (I.M.S.)	29/3/04 to 1905	Medical Officer.
Bartlett, H. T., Capt.	15/6/57 to 3/1/58	First Commandant, and officer who raised the Regiment. Was Cantt. Magte., Peshawar.
Basu, A. N., Capt. (I.M.S.)	19/1/21 to 15/4/21	
Bennett, O. D., Major	8/1/05 to date	D.C.O. ; Qr.-Mr., 1/3/08 ; Khaibar Rifles, 1912 ; Mohmand Militia, 1920 ; S. Waz. Scouts, 1922 ; Offg. I.O. Frontier Mil., 1922 ; D.C.C., 19/8/19 ; rejoined 1923 for Wana Column.
Bernard, E. H., Major (20th P.I.)	12/7/06 to 1908	Second-in-Command, 1907 ; retired, 1908.
Birch, A. J. C., Lt.-Col. (S.C.) (2/Gurkhas)	1/7/82 to 1883	Offg. Commandant.
Blake, W. J., Lieut. (21st Lancers and 2/66th Punjabis)	2/10/22 to date	Coy. Officer, Field Cashier, Wazir Force, 1922-23.
Blood, J., Surgeon	25/2/78 to 1878	—
Bond, W. J. H., Lieut. (2/W.I.R.)	11/7/80 to 1883	Offg. W.O. on probn., Q.M., 1882 ; Comst. Dept. on probn., 1883.
Bonniface, B. H., 2/Lieut. (I.A.R.O.)	21/10/18 to 1919	—
Branson, C. E. D., Capt. (Genl. List Infy.)	5/10/70 to 1874	Thomason College, Roorkee, Coolie Corps Column, Lushai, 1872.
Britton, T. R., Lt. (A./Capt). (I.A.T.C.)	8/3/19 to 1920	R.T.O., Kohat.
Brommage, J. C., Lieut. (I.A.R.O.)	3/9/18 to 1919	On probation.
Brown, F. H. J., 2/Lieut. (I.A.T.C.)	27/10/18 to 1919	

APPENDIX I—(*continued*)

NAME AND RANK	PERIOD SERVED	APPOINTMENTS, ETC.
Brown, L. C. H., Lieut.	10/10/19 to 1921	With 2/112th Infantry, 1921.
Bruce, H. M., Lieut. (54th Foot)	20/7/76 to 1877	Offg. on probation.
Burn, L. C. P., Lt.-Col. (33rd P.I.)	May, 1901 to Sept., 1901	Offg. Commandant.
Burrows, L. B., Lieut.	2/3/18 to 1918	—
Bwye, M. Lieut. (I.A.U.L.)	5/2/20 to date	Coy. Officer ; Offg. Q.M., 1922.
Cadell, —, Capt. (38th Dogras)	1896	Attached for Suakin Expedition, 1896.
Caird, W. E., Asst.-Surgeon	1867	Offg. Medical charge.
Calthorp, C. W., Asst.-Surgeon	18/3/70 to 1870	Doing duty.
Campbell, A. A. E., Brig.-Gen. (25th P.)	29/10/06 to 10/5/10	Commandant ; Comdg. Brigade, Quetta, 1913 to 1917 ; retired 1919.
Cargill, R. J., Major	24/8/03 to 1919	D.C.O. ; Malay States Guides, 1913 ; invalided from Mesopotamia, 1916 ; *Asst.-Comdt.* Prisoner-of-War Camp, 1919 ; retired, 1919.
Carleton, K. O., Lieut. (25th Punjabis)	Sept., 1916 to Jan., 1918	With Bn. in Mesopotamia.
Carpendale, P. M., Lt.-Col. (21st Punjabis)	15/12/00 to April, 1901	Offg. Commandant.
Carter, C. A. E. S., Lieut. (20th N.I.)	15/12/60 to 1862	Doing duty officer.
Chester, C. W. R., Col. (S.C.)	11/11/79 to 15/9/86	Commandant.
Chilcott, W. J., Lieut.	Aug., 1918 to June, 1919	With Bn. in Mesopotamia.
Close, J. K., M.D. Surg.-Capt (I.M.S.)	11/5/92 to 1897	Went to civil employ.
Collinson, W. J., M.B., Capt. (I.M.S.)	11/2/08 to 1910	Medical charge, Civil U.P., 1909 (tempy.) rejoined.

Conner, T. S., Capt. ... (73rd Carnatics)	... 12/5/22 to date	Coy. Officer and Offg. Coy. Comdr.
Copleston, R. T., Lieut. (I.A.R.O.)	... 23/9/18 to 1919	—
Cook, R., Lieut. ... (2/66th Punjabis)	... 6/10/22 to date	Coy. Officer, R.T.O., Khirgi, 1922–23.
Cookson, H., Surg.-Major	... 28/2/76 to 1876	Medical charge.
Cookson, S. M., Lieut. ... (21st Punjabis)	... 26/1/12 to 1913	Attached in China.
Cotton, J. W. M., Lieut. (21st Hussars)	... 18/12/68 to 1869 ...	Second Wing Subaltern.
Cox, A. D., Col. ... (21st Punjabis)	... 16/5/17 to 1921	Comdt., ; with Bn. in Mesopotamia, 1917 to 1919. Despatches.
Cox, F. A. D., Lieut. ... (17th N.I.)	... 7/1/61 to 1863	Doing duty officer.
Cragg, F. W., M.B., Lieut. (I.M.S.)	... 19/8/08 to 1909	Offg. Medical Officer.
Crampin, R. W., Capt. (2/66 Punjabis)	... 19/11/22 to 1923 ...	Attached ; Coy. Officer ; transferred to 1/15th Punjab Regiment, 1923.
Crichton, G. C. L., 2/Lieut. (D.L.I.)	... 31/10/22 to 12/8/23 ...	Coy. Officer ; transferred to 5/11th Sikh Regiment.
Crofts, A. M., Surg. ...	... 26/7/81 to 1882	Offg. Medical Officer and Civil, Jhelum.
Croghan, S. C., 2/Lieut. (I.A.R.O.)	... 21/10/18 to 1919 ...	—
Crookshank, A. C., Lieut. (35th Foot)	... 3/7/67 to 1868	2nd Wing Subaltern and Offg. Adjutant.
Cuningham, A. P., 2/Lieut.	... 1913 ...	Offg. with 127th Baluch. Infantry.
Cuninghame, T. B., Lieut. (I.A.R.O.)	... 5/9/15 to 1919	With Bn. in Mesopotamia, 30/12/15 to 20/3/19.
Currie, —, Asst.-Surg. ...	... Jan., 1858 to May, 1858	—
Darwell, — Capt. ...	... Aug., 1857 to Jan., 1858	Second-in-Command.
Davys, G. I., Lieut. (I.M.S.) ...	... 24/10/02 to 1903 ...	Attached ; Offg. Medical Charge.

APPENDIX I—(*continued*)

NAME AND RANK	PERIOD SERVED	APPOINTMENTS, ETC.
Dening, L., C.B., D.S.O., Maj.-Gen. (S.C.)	20/4/71 to 1899 … …	Second Wing Subtn.; Offg. Qr.Mr.; Offg. Bde. Major, Sialkot, 1881; Offg. S.S.O., Jhelum, 1882; Wing Comdt. 30/3/85; Second-in-Command, 1/5/87; D.S.O. for services in Burmah, 1886–87; Comdt., 11/5/94 to November, 1899. Comdg. Derajat Bde., 1901; Waziristan Blockade; C.B., 1903. Hon. Colonel of Regiment, 1903; Maj.-Gen. Comdg. Burmah Div., 1907; died 1910.
Dennys, W. A. B., Lt.-Col. …	27/3/83 to 1901 … …	Offg. Wing Officer, A.D.C. to L.G. Punjab, 1885; Adjutant, Agra Vol. Rifle Corps, 1892; transferred to 31st Punjabis and became Comdt. 1903.
Des Voeux, C. H., Maj. …	1896 … … …	Attached for Suakin Expedition; Offg. Comdt. *vice* Worlledge.
Dick, A., Capt. … … … (R.A.M.C.)	July, 1918 to Aug., 1918	With Bn. in Mesopotamia.
Dillon, G. F. H., C.B., Lt.-Col. (Suffolk R.) … … …	25/5/84 to 26/9/06	Offg. W.O.; Adjt., 1/5/87; F.S. Lushai, 1889; trsfd. to 40th Punjabis, 1890; D.A.A.G., Waz. F.F., 1894; D.A.A. and Q.M.G., Abbottabad Mov. Col., 1895; D.A.A. and Q.M.G., Malakand F.F., 1897; despatches (4 times); Brevet Maj., Staff, 1894–1899; returned as Comdt. 1899 till he died in 1906; Offg. D.A.C., India, 1900; A.A.G., Lahore, 1901; C.B., 1903.
Dillon, H. C. W., 2/Lieut. (son of Lt.-Col. G. F. H. Dillon)	14/3/10 to 1915 … …	Served with 24th Punjabis in Egypt and Mesopotamia; killed in action, Mesopotamia, 1915.

Name	Period	Remarks
Dodsworth, W. F., Capt. (Gen. List Infy.)	25/5/71 to 1872	Thomason College, Roorkee, 1872.
Douglas, H. McD. De W., Capt. (Staff Corps)	11/1/68 to 1872	Offg. W. O.
Driver, R. A., Capt. (2/19th Punjabis)	17/5/21 to 30/9/22	Coy. Officer, Offg. Coy. Comdr. ; retired.
Drysdale, H. D., Lieut.	14/11/08 to 1915	Qr.Mr., 1913 ; killed in action in France, whilst serving with the 11th Bn. Royal Scots, on 31/8/15.
Dunsterville, L. C., C.B., Maj.-Gen. (20th P.I.)	1898 and 1904	Attached as Offg. W.C., 1898 ; Second-in-Command and Offg. Comdt., 1904 ; returned to 20th P.I., which he commanded ; conducting officer, France, 1914-15 ; later Bde. Comdr., Peshawar ; Comdr. of " Dunster Force " in Persia, 1917–1918.
Eastmond, R.C., (I.A.R.)	25/8/15 to 1916	With Bn. in Mesopotamia, 30/12/15 to 21/4/16
Edwards, C. A., Lieut. (R.W.F.)	28/10/87 to 1888	Attached as Offg. W.O. on probation.
Edwards, E., Major (29th Punjabis)	14/1/22 to date	Offg. Second-in-Command, ; Pensions Investigating Officer, 1923.
Edwards, M. F., Capt. (1/19th Punjabis)	23/1/21 to 30/9/22	Coy. Officer ; Offg. Coy. Comdr. ; retired to Natal.
Elliott, P. W., Maj. (1/19th Punjabis) (20th Infy.)	—	Attached, 1912, in China.
Fagan, C. H., Lieut. (36th N.I.)	13/1/59 to 1859	—
Fagan, —, Capt. (1st P.I.)	1896	Attached for Suakin Expedition.

APPENDIX I—*(continued)*

NAME AND RANK	PERIOD SERVED	APPOINTMENTS, ETC.
Farwell, G. A. L., M.C., Capt. (son of Lt.-Col. W. C. Farwell)	16/4/15 to date	With Bn. in Mesopotamia, 30/12/15 to 14/10/19 ; Adjutant, 1917 to 1922 (mention) ; M.C. ; Waziristan, 1921–22 (mention) ; Adjt., B. & N.W. Railway (Auxiliary Force), Gorakhpur, 1922.
Farwell, W. C., Lt.-Col. (G.L.I.)	28/12/85 to 11/5/94	Second-in-Command ; Comdt., 1/5/87.
Field, C. W., Capt. (S. Wales Borderers)	11/6/86 to 1895	Wing Officer ; Q.M., 22/1/90 ; Offg. Cantt. Magte., Multan, 1891 ; A.J.A.J., Quetta, 1896.
FitzGerald, A., Lt.-Col. (Gen. List)	19/7/61 to June, 1886	Doing Duty Officer ; Offg. Adjt., 1861 ; Adjt. 13/5/65 to 1876 ; Abyssinian Expeditionary Force, 1868 ; Offg. Second-in-Command, 1873 ; Bde. Major, Mooltan, 1879 ; Bde. Major, Quetta. 1882 ; Brevet Lt.-Col., 1884 ; Comdt., 17th B.I., 1886 ; Offg. Comdt., Burmah, 1886.
FitzGerald, O. A. G., Col.	6/1/97 to 1897	Supy. attached 20th P.I. Malakand, 1897 ; trsfd. to 18th B.L. ; later drowned with Lord Kitchener, 1916.
Fitzherbert, G. A., Lieut. (I.A.R.)	Mar., 1918 to June, 1918	With Bn. in Mesopotamia.
Fitzpatrick, B. P. S., Lieut.	1922 to date	Inspector Physical Training, Peshawar, 1922.
Flemin, J. McN., M.D., Bde. Surg. (I.M.S.)	30/11/82 to 1892	With 4th B.C., 1891.
Forster, G. H., Lieut. (Gen. List Infy.)	14/9/66 to 1867	—
Fry, A. B., Capt. (I.M.S.)	26/6/03 to 1904	Medical charge.

Name	Dates	Remarks
Fulford, C. J. R., Capt. (S.C.)	13/9/70 to 1882 ...	Second Wing Subaltern ; Qr.Mr., 28/3/71 ; Staff College, 1879–80 ; Q.M.Gs. Dept., 1882.
Fulton, J. D., M.C., Capt.	17/11/09 to date	D.C. Officer ; Adjutant, 3/12/14 ; Mesopotamia, 1916–20 ; Staff Capt., 36th Bde., 1917 ; B.M., 55th Bde., 1920 (6 mentions and M.C.) ; Coy. Comdr. ; rejoined Bn., Waziristan, 1922.
Gardiner, T., Lieut. (H.M. 98th Foot)	24/2/59 to Dec., 1859	Adjutant.
Geary, H. V., M.C., Capt. (2/69th Punjabis)	25/3/21 to 14/11/22	Coy. Officer, Staff Capt., 7th Bde., Tochi, 1922 ; retired, 1922.
Gerard, A. F., Lieut. (Late 3rd E.R.)	19/10/64 to 1866	—
Gilbert, C. E. L., Lt.-Col. (I.M.S.)	1912	Medical Officer.
Giles, W., Maj. (20th P.I.)	2/4/05 to 1906	Second-in-Command ; but never joined.
Gill, C. A., Col. (I.M.S.)	5/5/04 to 1905	Transferred to Civil.
Glover, —, Asst.Surg	Nov., 1857 to Dec., 1857	—
Goddard, A. G., I.A., Capt. (2/66th Punjabis)	2/10/22 to date	Coy. Officer and Offg. Coy. Comdr.
Gowan, B. E., Lt.-Col. (S.C.)	15/9/86 to 1/5/87	Commandant.
Graham, C. T., Capt. (I.A.R.O.)	12/11/15 to 1919	—
Grainger, T. A., M.B., Surg.	12/3/96 to 1896	Offg. Medical Charge.
Grant, —, Capt. (2/4th Gurkhas)	May, 1896 to Dec., 1896	Attached for Suakin Expedition.
Graves, S. H. P., Maj. (S.C.)	10/12/77 to 1891	Adjutant, 14/5/78 ; Offg. Dist. Suptd. Police, Port Blair, 1883 ; Dy.I.G. Police, Burmah, 1888 ; transferred to 40th Punjabis.
Gray, G. E., Lieut.	12/12/22 to 18/4/23	Coy. Officer ; attd. 91st Punjabis for Wana Column, 1923 ; trsfd. to 2/14th Punjab Regt.

APPENDIX I—(*continued*)

NAME AND RANK	PERIOD SERVED	APPOINTMENTS, ETC.
Green, —, Lieut.	1858	Attached.
Green, W. G., Lieut.	1922 to 1923	Attached, but never joined ; Cantt. Magte., D.I.K.
Grey, A. J. H., Maj. (U.L.)	10/10/99 to 1917	Offg. W.O. ; to China with 20th and 21st Punjabis, also attached 30th Punjabis, 1905 ; Pol. employ, 1907.
Grove, H. M., 2/Lieut. (Devon Regt.)	1/11/88 to 1889	Offg. W.O. on probation.
Gurdon, P. R. T., Lieut. (S.C.)	1/7/85 to 1886	Depot, Meerut.
Haldane, E. H. V., Lieut. (Connaught Rangers)	19/10/82 to 1883	Offg. W.O. on probation.
Hall, G. C., Surg.	12/5/73 to 1874	Offg. in Medical Charge.
Hancock, F. H., Lt.-Col. (Cheshire Regt.)	22/8/81 to 1902	Offg. Wing Officer ; Qr.Mr., 1/7/82 to 1889 ; Wing Comdr., 1892 ; Second-in-Command, 4/4/00 ; Comdt., 87th Punjabis, 1902.
Harington, H. A., Lieut. (Manchester Regt.)	15/12/92 to 29/9/97	Wing Officer ; with 25th Punjabis, 1896 ; A.R.O., Pathans, 1897 ; killed, Malakand, 29/9/97.
Harrison, J. B., M.D., Surg.-Maj.	29/6/67 to 1869	Medical Charge, 1869 ; Medical Storekeeper, Sialkot.
Hartle, J. C., Lieut.	17/8/17 to 1918	—
Hassan, H., Surg.	17/12/84 to 1887	Offg. Medical Charge.
Haynes, W. F., Lieut. (21st Punjabis)	26/1/12 to 1912	Attached in China.
Heathcote, F. C., Lieut. (I.A.R.)	Nov., 1916 to June, 1919	With Bn. in Mesopotamia.
Hepburn, A. B., Maj. (32nd N.I.)	14/7/65 to 1884	Doing Duty Officer ; Wing Subaltern ; Qr.Mr., 10/2/69 ; Offg. Adjt., 1870 ; Wing Officer, 1/4/76 ; Wing Comdr., 14/5/78.

Name	Period	Remarks
Hilliard, W. A., Lieut. (I.A.R.)	May, 1918 to Mar., 1919	With Bn. in Mesopotamia.
Hilson, A. H., M.D., Asst.-Surg.	16/9/58 to 1866	Civil Moradabad and Gorakhpur, 1863 and 1864.
Hodge, E. H. V., Lieut. (I.M.S.)	15/7/11 to 1914	Offg. Medical Charge.
Hodgson, B. P., Lieut. (Gen. List)	11/2/60 to 1868	Doing Duty Officer ; Q.M., 20/1/64
Hodgson, C. N., Lieut. (Late 10th N.I.)	11/2/60 to 1869	Doing Duty Officer ; Q.M., 20/1/64.
Hourihane, T. C., Lieut.	1919 to 1920	At Depot, Bareilly, 1919 ; killed on 1/2/20, with 4/39th Garhwal Rifles.
Howell, V., Lieut. (87th Punjabis)	Sept., 1919 to Oct., 1919	With Bn. in Mesopotamia.
Hughes, T. A., M.B., Capt. (I.M.S.)	11/3/15 to 1916	—
Hughes, T., Lieut. (Welch Regt.)	3/6/18 to 1922	Coy. Officer at Depot, 1921 ; retired, 1922; and joined I.C.S.
Hugo, J. H., M.B., D.S.O., Col. (I.M.S.)	23/3/00 to 1902	Transferred to Civil ; Civil Surgeon, Kashmir, 1922.
Hunt, J. M., Lieut. (87th Punjabis)	Aug., 1916 to Jan., 1918	Attached, in Mesopotamia.
Hunter, C., Lieut. (H.Ms. 81st Foot)	30/6/58 to 1859	Adjutant.
Hunter, G. Y. C., Surg.-Capt.	1898 —	Offg. Medical Charge.
Husband, G. S., M.B., Capt. (I.M.S.)	9/6/09 to 1911	Offg. Medical Charge.
Ivens, H. T. C., Maj. (7th R.F.)	16/1/02 to date	D.C.O. ; Wounded Mohmand Expedition, 1908 ; N. Waz. Militia, 1910 ; wounded 25/1/17 in Dahra Bend (Tigris) ; Offg. Comdt., short periods, Mesopotamia, 1917–18 ; G.S.O., Peshawar, Afghan War, 1919 ; R.S.O., Jullundur, 1921 ; Second-in-Command.

APPENDIX I—(continued)

NAME AND RANK	PERIOD SERVED	APPOINTMENTS, ETC.
James, A. A., Lieut. (129th Punjabis)	20/8/97 to 1898	Attached ; Offg. W.O., temporary.
Jardine, A., Capt.	1/6/22 to 1923	Offg. Coy. Comdr. ; S.S.O., Kotkai, 1922 ; Staff Capt., 21st Bde., 1923 ; trsfd. to 25th Punjabis, 1923.
Jee, N. B., Lieut.	16/1/19 to 1920	M.G. Corps, I.A., on probation.
Jessop, F. T., Lieut. (I.A.R.)	27/4/17 to 1919	With Bn. in Mesopotamia, 18/2/19 to Oct., 1919.
Johnstone, J. W., M.D., Surg.-Maj.	13/8/81 to 1883	Offg. 9th B.C., 1883.
Kapur, K. I., Lieut. (I.M.S.)	June, 1916 to Aug., 1916	With Bn. in Mesopotamia.
Ker, M. A., Lt.-Col. (I.M.S.)	4/10/10 to 1911	With 2/5th G.R.
Kilkelly, C., M.B., Asst.-Surg.	26/2/61 to 1861	Civil, Aligarh.
King, H., Capt. (89th N.I.) (I.A.U.L.)	25/5/58 to 29/10/61	Second-in-Command, 13/1/59 ; left on appointment as Offg. Comdt., 13th N. Infy.
Kidd, F. S., 2/Lieut.	28/12/18 to 1919	Attached.
Lapsley, J. B., M.B. Capt· (I.M.S.)	12/11/12 to 1918	Medical Officer ; with 86th Infy., 1914 ; also 1917 and 1918.
Lane, C. M., Lieut.	5/9/18 to date	Coy. Officer ; seconded to " B " Divn. Signals, 1922.
Lawson, O. H., Maj. (North'd Fus.)	18/9/92 to 11/3/16	Offg. W.O. ; Qr.Mr., 11/11/99 ; D.C.C., 23/11/00 ; Transport Officer, Sikkim, Tibet Mission, 1904 ; mortally wounded, 8/3/16, at Dujailah, and died, 11/3/16, at Amarah, on Tigris.

Name	Dates	Remarks
Lees, L. H., M.B., Surg.	10/2/70 to 1870	Medical Charge ; Civil, Ambala.
Lewis, T. L. (11th Foot)	24/1/70 to 1879	First Wing Subaltern ; Qr.Mr., 28/3/71 ; Offg. Adjt., 1873 ; Adjt., 1877 ; Army Comd. Dept. (probation).
Lewis, T. M., Capt. (T.C.)	23/1/21 to 17/8/22	Coy. Officer ; retired.
Little, I. M., Capt. (Royal Marines and Sussex Regt.)	27/12/00 to 1916	D.C.O. ; Qr.Mr., 1/10/04 ; Cantt. Magte., 1909 ; Depot, 1912 ; D.C.C., 10/5/12 ; wounded, Dujailah, 8/3/16 ; invalided.
Lockhart, Sir W. S. A., Gen. (44th N.I.)	1861 to 1862	Doing Duty Officer and i/c Adjutant's Office and Station Staff ; later Sir William Lockhart, K.C.B., etc., C.-in-C., in India.
London, W. C., Lieut. (W. Yorks Regt.)	17/11/82 to 1884	Offg. W.O. on probation.
Longmore, C. M., Lt.-Col. (S.C.)	8/5/63 to 1876	Second-in-Command, 20/1/64 ; Offg. Comdt., 18/2/75 to 1876.
Lyall, J. H., Lieut. (I.A.R.O.)	1916 to May, 1918	Served with the Bn. in Mesopotamia. Departed with a Company transferred from Mesopotamia to 1/152nd in Palestine.
Macaulay, T. G., Ensign (Gen. List)	23/2/61	Doing duty (not joined).
MacCartie, J. FitzG., Lieut. (Durham L.I.)	Mar., 1886 to 1887	Offg. W.O. ; killed in Burmah.
MacInnes, I. L., M.B., Lieut.	4/11/00 to 1901	For temporary duty.
Maclaren, W. de B., Capt. (2/25th Punjabis)	1922 —	Attached to Depot, Jullundur.
Macwatt, R. C., M.B., Surg.	26/5/90 to 1891	Offg. Medical Charge ; F.S. Hazara.
Maidman, —, Lieut.	Oct., 1858 to Jan., 1859	—
Malcomson, G. E., M.D., Capt. (I.M.S.)	22/2/15 to 1917	Medical Officer ; with Bn. in Mesopotamia, 30/12/15 to 18/6/16.
Manley, W. J., Capt. (2/66th Punjabis)	1922	Posted May, 1922, but retired at once and never joined.
Marett, J. R., Lieut. (Staff Corps)	8/12/66 to 1867	Second Wing Subaltern ; Offg. Qr.Mr.

APPENDIX I—(*continued*)

NAME AND RANK	PERIOD SERVED	APPOINTMENTS, ETC.
Martin, J. E. L., Lieut. (R.A.)	8/9/21 to date	Coy. Officer.
Maude, E. A., D.S.O., Lt.-Col.	17/1/04 to Dec.. 1923 ...	D.C.O. ; Acting Comdt. in Mesopotamia, 1917 (Acting Lt.-Col.) ; D.S.O. ; B.M., 12th Bde., M.E.F., 1917 (mention) ; D.A.A.G., Army H.Qrs., Simla, 1919 ; again Acting Comdt., Waziristan, 1921 (Acting Lt.-Col.) ; (mention) ; trsfd. to 25th Punjabis as Comdt., Dec., 1923.
Meacham, W. M., ... (41st Foot)	24/8/69 to 1870	Second Wing Subaltern (on probation).
McConnell, J. K., Capt. (2/66th Punjabis)	31/8/22 to 1923	Attached, Coy. Officer ; trsfd. to 1/66th Punjabis.
McKenna, C. J., Asst.-Surg. ...	6/6/71 to 1873	Medical Charge.
Milne, R., Lieut. ... ... (15th N.I.)	8/1/61 to 1861	Doing Duty Officer.
Morris, E. W., D.S.O., Capt. ... (Connaught Rangers)	1921	Comdg. Labour Corps, Waziristan, 1922.
Newington, C. D. G., Lieut. ... (Genl. List Infy.)		No date of joining ; in Army List, 1866.
Nicholas, J. W., ... (I.A.R.O.)	18/2/17 to 1917	With Bn. in Mesopotamia, Jan., 1917, to March, 1917.
Nield, R., I.C.S., Lieut. ... (I.A.R.O.)	May, 1916 to Mar., 1917	With Bn. in Mesopotamia, 1916 and 1917 ; Intelligence duty, G.S., Mesopotamia, 1917–18 ; I.C.S.
O'Bryen, C. W., Maj. ... ... (27th P.I.)	15/6/03 to 1903	Second-in-Command, with 27th Punjabis, (tempy.).
Ogilvie, W. H., Surg.-Capt. ... (I.M.S.)	1897 ...	—
O'Keefe, D. S. A., M.B., ... (I.M.S.)	8/8/06 to 1907	Medical Charge.

Name	Period	Remarks
O'Malley, —, Lieut. ...	Aug., 1857 to May, 1858	Doing Duty Officer.
Parminter, A. S. P., Lieut. ...	13/12/17 to 1918 ...	—
Perkins, R. H., Surg. ...	29/12/66 to 1867 ...	—
Perry, A. H. S., Lieut. (I.A.R.O.)	27/9/17 to 1919 ...	With Bn. in Mesopotamia, Oct., 1916 to May, 1917.
Peyton, J., Lieut. ...	15/6/57 to Aug., 1857 ...	First Second-in-Command.
Pinniger, R. C., Lieut. (I.A.R.O.)	25/8/15 to 1917 ...	With Bn. in Mesopotamia, 30/12/15 to 4/6/16.
Pollock, A. H., M.C., Capt. (2/26th Punjabis)	22/1/22 to date ...	Coy. Officer ; Offg. Coy. Comdr.
Powell, —, Lieut. ... (23rd Pioneers)	1896	Attached for Suakin Expedition.
Raikes, R. F., Lieut. ...	23/8/22 to date ...	Coy. Officer ; seconded to " B " Signals Divn., 1922.
Rankin, H. H., Lieut. ... (Gen. List Infy.)	20/2/66 to 1867 ...	Second Doing Duty Officer (on probation); leave to Europe.
Raschen, G. H., Lieut. (I.A.R.O.)	April, 1916 to July, 1916	With Bn. in Mesopotamia.
Rattray, C., Brig.-Gen. p.s.c.	26/8/91 to 1912 ...	Qr.Mr., 1891 to 1897 ; Adjt., 25/8/97 ; Staff College, 1901–2 ; Offg. B.M., Derajat, 1907 ; D.A.A.G., Derajat, 1909 ; G.S.O., (2nd Grade), 1910 ; to 20th Punjabis, 1912, as Comdt. ; served on Staff of M.E.F., as D.A. & Q.M.G. ; C.B. and Brevet Col.
Ravenhill, M. H., 2/Lieut. (I.A.R.O.)	5/9/15 to 1917 ...	—
Ravenshaw, H. A., Maj. (73rd F.)	31/8/77 to 4/4/00 ...	Offg. W.O. (on probation) ; Offg. Q.M., 1879 ; Adjt., 21/10/80 ; Second-in-Command, 11/5/92.
Raynor, V. H., Lieut. ... (87th Punjabis)	Jan., 1917 to June, 1917	With Bn. in Mesopotamia.
Read, A. W. C., Capt. ... (H.Ms. 51st Foot)	13/2/62 to 1863 ...	Offg. Second-in-Command.

I

APPENDIX I—(*continued*)

NAME AND RANK	PERIOD SERVED	APPOINTMENTS, ETC.
Reade, B. E., Lieut. ... (Late 67th N.I.)	22/3/64 to 1866 ...	Comdg. Depot, Gorakhpur, 1866.
Reed, A. K., Surg.-Maj.	18/12/76 to 1877 ...	—
Ridgway, R. T. I., C.B., Brig.-Gen. (Connaught Rangers)	3/3/89 to 1891 ...	Offg. W.O. ; trsfd. to 40th Pathans ; later Comdt. 33rd P., and Brig.-Gen. Comdg., Kohat Bde.
Roberts, T. L., Lieut. ... (H.Ms. 87th Foot)	15/6/57 to Jan., 1859 ...	First Adjutant.
Robertson, H. L. C., Capt. ...	3/11/21 to date ...	Coy. Officer ; Adjt., 30/10/22 ; with Bn. in Mesopotamia, Oct., 1917 to Oct., 1919 ; Chevalier of Crown of Rumania ; Waziristan, 1921.
Rodriques, F. ... (I.M.S.)	Sept., 1917 to June, 1918	With Bn. in Mesopotamia.
Rollinson, R. G., Lieut. (I.A.R.O.)	22/5/19 to 1920 ...	With Bn. in Mesopotamia, May, 1918 to April, 1919.
Rootes, A. H., Capt. ...	1/11/19 to date ...	Coy. Officer ; Waziristan, 1921 (mention).
Ryland, H. G., Lieut. ... (S.C.)	4/10/78 to 1879 ...	Offg. Wing Officer.
Salusbury, R. T. G., Maj. ...	5/12/03 to 22/1/23	D.C.O. ; Adjt.,1/4/08 ; with 120th R., 1914 ; wounded at Shaiba (Mesopotamia) ; Staff appointments 1916 to 1921 ; D.A.Q.M.G., Kohat, and A.H.Q. ; retired, 1923.
Saxton, P. D., Capt. ... (20th P.I.)	1912 and 1916 ...	Attached 1912–13, in China, and again in 1916, in Mesopotamia.
Scott-Moncrieff, W. E., M.B., Capt. (I.M.S.)	14/9/98 to 1899 ...	Offg. Medical Officer.
Sevenoaks, P. L., 2/Lieut. ... (I.A.R.O.)	25/8/18 to 1919 ...	—

Name	Date	Remarks
Shearer, J. E., M.C., Capt.	3/11/11 to date	With 20th Infy., 1912 ; Offg. Qr.Mr., 1915, and Qr.-Mr., 1916 and 1917 (mention) ; M.C. in Mesopotamia ; with 4/39th Garhwal Rifles, Waziristan, 1920 ; Comdg. Depot, 1921.
Shuttleworth, B. W., Maj.	24/6/03 to 1903	D.C. Officer.
Singh, B. J., Surg. (I.M.S.)	1891 to 1892	—
Smith, A. S., Lt.-Col. (Beng. Infy.)	23/6/69 to April, 1871	Offg. Commandant.
Smith, A. W., Capt. (R.A.M.C.)	Aug., 1918 to Oct., 1919	With Bn. in Mesopotamia.
Smith, M. G., Lt.-Col. (59th N.I.)	20/2/64 to Sept., 1881	Wing Officer ; Second-in-Command, 1/4/76 ; Comdt., 7/8/77.
Steer, J. M. B., Capt.	25/4/21 to date	Coy. Officer ; Offg. Coy. Comdr., 1921 ; G.S.O.3, Wazir Force, 1923.
Stephenson, J., M.B., Surg.-Lt.	6/2/97 to 1897	Offg. Medical Charge ; Plague duty.
Stevenson, J. H. W., 2/Lieut. (46th P.)	Oct., 1916 to Feb., 1917	With Bn. in Mesopotamia ; mortally wounded, Hai Bridge-head, 5/12/17.
Stewart, J. L., M.D., Asst.-Surg.	30/6/58 to Jan., 1859	—
Stone, B. S., Lieut. (N. Staff. Regt.)	19/3/09 to 1910	D.C. Officer.
Stoney, P. S., *p.s.c.*, Lt.-Col. (U.L.)	22/4/00 to date	Offg. W.O. ; A.R.O. Pathans, 1903 ; Q.M., 1906 ; Staff College, 1909 and 1910 ; G.S., Br.A.H.Q., 1911–16 ; M.S., Br.A.H.Q., 1916 ; G.S., M.E.F., 1917–18 ; (mention) ; Comdt. (tempy.), 2/66th P., 1918–21 ; Comdt., 26th P., 8/2/22 ; Waziristan, 1922–23.
St. Ruth, G. H. B., M.C., Capt. (37th Cavalry)	24/6/21 to 3/6/22	Coy. Officer and Offg. Coy. Comdr. ; retired on reduction.
Tarr, W., M.B., Capt. (I.M.S.)	4/1/05 to 1906	Medical Officer.
Teape, C. C. T., Lieut.	1917 to 1920	With Bn. in Mesopotamia, Dec., 1917 to Dec., 1918 ; died in Peshawar, 1920.

NAME AND RANK	PERIOD SERVED	APPOINTMENTS, ETC.
Thomas, J. E., Capt. ...	26/1/18 to 16/4/22	Coy. Officer ; Staff Capt., 28th Bde. ; Staff Capt., 21st Bde., 1921, Waziristan (mention in despatches) ; retired on reduction.
Thomas, R. E. N., Lieut. (2nd Bn. 9th Foot)	18/7/77 to 1878 ...	Offg. W.O. (on probation).
Thompson, I. F. R., Lt.-Col. ... (Hants Regt.)	26/7/91 to 25/1/17	Offg. W.O. ; with 20th P.I., 1894–95 (Wana) ; Q.M., 10/12/97 ; Tpt. Regn. Officer, 1903 ; D.C., 1/5/00 ; Second-in-Command, 10/5/12 ; Acting Comdt., 1916, Mesopotamia ; killed while leading the Regiment in Dahra Bend, 26/1/17.
Torrie, L. J., Capt. ...	20/9/09 to 1920 ...	D.C. Officer ; Posted to 12th Royal Scots from 21/8/14, and served in France ; severely wounded, and later invalided.
Tulloch, A., Col. (S.C.)	7/8/77 to 1885 ...	Second-in-Command.
Turnbull, G. O., D.S.O., Maj.	8/4/99 to 1918 ...	Offg. W.O. ; Qr.Mr. 25/5/01 ; Adjt., 1/4/04 ; D.C.C., 14/7/09 ; Attd. Royal Scots Fus., France, 1914 ; D.S.O. ; wounded severely ; attd. G.S. Br. A.H.Q., Simla, 1917 ; A.A.G., Palestine, Expeditionary Force, 1918 ; Second-in-Command, 16/4/18 ; retired, 22/1/22, owing to wound disability.
Vallings, J., Lieut. ...	Dec., 1859 to Jan., 1860	Acting Adjutant.
Van der Gucht, T. E., Lieut. ... (5th N.I.)	31/12/57 to May, 1858 ...	—
Veale, T. S., Asst.-Surg. ...	24/4/66 to 1867 ...	In Medical Charge, Simla M,C.
Vincent, M. V., Lieut. ...	23/11/21 to 17/5/23	Attached, but never joined ; M.F.O., Sorarogha.
Waghorn, A. R., M.D., Surg.-Maj.	24/11/74 to 1875	Medical Charge.

Walton, G., Lt.-Col. ... (R. War. Regt.)	...	3/11/94 to 1904 ...	... Offg. W.O. ; Adjt., 4/4/00 ; trsfd. to 46th P., 1904 ; Adjt., Simla Volunteers, 1908 ; D.A.A.G., Lahore Divn., 1914 ; and later A.Q.M.G., ; Brevet Lt.-Col. Comdt., 46th P.
Walton, L. B., Brig.-Gen. (Cheshire Regt.)	...	17/10/87 to 1917	... Offg. W.O. ; Adjt., 21/1/91 ; Comdt. and D.S.P., Port Blair, 1898 ; D.C.C., 1/5/00 ; returned 1904 ; Second-in-Command, 1908 ; Comdt., 10/5/12 ; Comdg. Bn. in Meso-potamia, 1916 ; Acting Bde. Comdr. on Tigris ; invalided and died, 1917 ; Brevet Col.
Warren, D. S., Capt. ... (I.A.R.O.)	...	7/9/15 to 1919 ...	... With Bn. in Mesopotamia, Aug,, 1916, to Dec., 1918 ; Acting Qr.Mr., 1917 and 1918 ; Staff Capt., " Percoms," 1918 (mention).
Wauchope, R. C., Lt.-Col.	...	30/5/73 ...	... —
West, A. C. E., 2/Lieut. (47th Sikhs)	...	12/3/23 to 1923 ...	... Coy. Officer ; trsfd. to 2/10th Baluch R.
White, A., M.D., Surg.	...	12/11/58 to 15/9/59	... —
Whitehead, A. W. N., 2/Lieut. (I.A.U.L.)		23/12/18 to 1919	... Attached.
Williams, B. F., Capt. ...	...	9/6/21 to 19/6/22	... Attached from Cavalry ; retired.
Williamson, C. I. H., Lieut. (S.C.)	...	30/8/87 to 1892 ...	... Wing Officer ; Burmah M.P., 1891.
Williamson, J., Col. ... (49th N.I.)	...	31/12/57 to 9/4/77	... Appointed Commandant six months after the Regiment was raised, and commanded it for nearly twenty years.
Wilson, C. G., Lieut. ... (21st P.)	...	5/4/22 to date ...	... Coy. Officer, attached.
Wiseman, D. J. C., Lieut.	...	7/4/21 to date ...	... Qr.Mr., 1921.
Worlledge, —, Maj. ... (35th Sikhs)	...	1896	... Attached for Suakin Expedition, 1896 ; Offg. Comdt. vice Dening (sick).
Young, H. L., Lieut. ... (Gen. List Infy.)	...	2/11/66 to 1867 ...	... Offg. First Wing Subaltern.
Young, J. W. F., Lieut.	...	19/5/21 to 24/7/22	... Coy. Officer ; retired.

APPENDIX II.

INSPECTION REPORTS.

PLACE.	DATE.	NAME OF INSPECTING OFFICER AND REMARKS.
Gonda	21/9/59	Brigadier Holditch, C.B., expressed his satisfaction.
	Jan., 1860	Brigadier Holditch, C.B., again inspected and reported favourably.
Meerut	7/3/63	His Excellency the Commander-in-Chief, Sir Hugh Rose, G.C.B., expressed his satisfaction at the way in which the Regiment had drilled on the occasion of presenting colours to the Regiment.
Meerut	2/3/85	His Royal Highness the Duke of Connaught, Commanding Meerut Division.
Meerut	15/3/88	Major-General Sir George Greaves, K.C.B., Commanding at Meerut, expressed his satisfaction.
Meerut	15/3/88	His Excellency the Commander-in-Chief expressed his satisfaction with the Regiment and with its general appearance.
Peshawar	1890	Brigadier-General F. J. Keen, C.B., Commanding Peshawar District, expressed his complete satisfaction.
Peshawar	Dec., 1890	His Excellency the Commander-in-Chief was much pleased and remarked :— " It is composed of a fine body of men, well drilled and turned out, and has been steadily improving in musketry in the last three years, and the figure of merit increased 22.81 points since last year." Major Dening was especially mentioned for the interest he took in musketry.
Peshawar	1891	Brigadier-General F. J. Keen, C.B., Commanding Peshawar District, expressed his complete satisfaction.
Peshawar	18/1/92	Brigadier-General F. J. Keen, C.B., Commanding Peshawar District, expressed his general satisfaction and noticed a considerable improvement in drill and steadiness.
Jhelum	9/3/93	Major-General Sir W. K. Elles, K.C.B., Commanding Rawalpindi District. His Excellency the Commander-in-Chief's remarks on this inspection were :—" The Regiment is in good condition and fit for service."

Jhelum	...	24/2/94	...	Major-General Sir W. K. Elles, K.C.B.
D.I. Khan	...	23/2/95	...	Colonel A. G. Hammond, V.C., C.B., A.D.C., D.S.O., temporary Commanding P. F. F. Commander-in-Chief's report on inspection. This report shows the 26th Punjab Infantry to be a very efficient regiment, ably commanded by Lieutenant-Colonel Dening. As regards musketry this regiment contains some of the finest marksmen in the Army.
Rawalpindi	...	11/3/96	...	Major-General Moorsom, Commanding Rawalpindi District, expressed himself exceedingly pleased with his inspection. Lieutenant-General Commanding :—A most satisfactory report in every respect, very creditable to Colonel Dening and all ranks of the Regiment.
Jullundur	...	5/2/97	...	Major-General Sir George Wolseley, K.C.B., Commanding Lahore District. The Lieutenant-General Commanding made the following remarks on this inspection :—" The 26th Punjab Infantry continues to maintain its high standard of efficiency, and its condition is in every way satisfactory, especially as regards its performance in musketry, which shows a most creditable improvement, notwithstanding the absence of the Regiment for some months on Field service at Suakin, and the consequent want of practice." The Commander-in-Chief's remarks were as follows :—" A very satisfactory report, reflecting great credit on the Commanding Officer."
Jullundur	...	31/3/98 (Left Wing only)	...	Major-General H. M. Evans, C.B., Commanding Lahore District. Commander-in-Chief's remarks :—" A very satisfactory report, and the conduct of the Afridi companies redounds to the credit of the Regiment."
Jullundur	...	15/3/99	...	Major-General G. Morton, K.C.I.E., Commanding Lahore District, said he was well satisfied with the condition of this Regiment, the men are thoroughly well trained, the interior economy is all that it should be, its efficiency reflects great credit on Colonel Dening. The Commander-in-Chief, Sir William Lockhart :—" I am very glad to find a regiment in which I have long taken so deep an interest so highly reported on."

APPENDIX II—*(continued)*

PLACE.	DATE.	NAME OF INSPECTING OFFICER AND REMARKS.
Jullundur ...	5/1/00 ...	Major-General Sir Gerald Morton, K.C.I.E., C.B., Commanding Lahore District, expressed himself well satisfied with everything he had seen.
Peshawar ...	14/1/02 ...	Brigadier-General Stratford Collins, Commanding Peshawar District. The remarks of the Lieutenant-General Commanding and His Excellency the Commander-in-Chief in India were satisfactory.
Peshawar ...	6/1/03 ...	Brigadier-General Stratford Collins, Commanding Peshawar District :—" Thoroughly satisfied with the conditions and appearance of the Regiment, especially the high state of efficiency in practical training."
Peshawar ...	5/1/04 ...	Major-General Sir E. Barrow, Commanding Peshawar District, reported well. Lieutenant-General Commanding :—" A satisfactory report, the high state of efficiency attained in signalling is very creditable."
D. I. Khan ...	Jan., 1905...	Lord Kitchener's test. General Officer Commanding, General Pearson, said :—" I have a very high opinion of the 26th Punjabis, a good tone prevails throughout the Regiment. There are a good lot of Native Officers in the Regiment, and non-commissioned officers and men are keen and of excellent fighting material." Lieutenant-General Commanding :—" The 26th Punjabis have been kept well up to their usual high standard during this past year." Commander-in-Chief :—" An excellent report in all respects."
D. I. Khan ...	Jan., 1906...	General Officer Commanding. Colonel Anderson :—" The Battalion is in first rate order. The whole personnel of the Regiment is of good quality and fine fighting material. The discipline is strict and good. The Regiment has, I consider, attained a very high standard of military efficiency indeed, and the greatest credit is due to Lieutenant-Colonel Dillon and his officers for the excellent results of their work."
D. I. Khan ...	Feb., 1907...	Major-General Adams, V.C., C.B., General Officer Commanding :—" A very efficient corps ; there is a good tone in the Regiment." Lieutenant-General Commanding :—" A very satisfactory report." Commander-in-Chief :—" Very satisfactory ; I am glad to see this Battalion is very efficient in musketry and signalling."

Dabbra (Near Tank, Derajat) ...	... Jan., 1908...	Major-General R. B. Adams, V.C., C.B., gave a very satisfactory report, and concluded :—" Very efficient, well trained, and thoroughly fit for active service." On April 18th His Excellency Lord Kitchener inspected the Regiment, at Kohat, and expressed himself thoroughly satisfied that it was keeping up its old reputation.
Kohat	... 5/3/09 ...	Lieutenant-General Sir J. H. Wodehouse, K.C.B., Commanding Northern Army, reported :—" A particularly fine regiment, good in the field, and with a remarkably fine soldierlike bearing on parade. The Regiment struck me as among the very best I have seen. Fire discipline and control good."
Kohat	... 1910 ...	Brigadier-General W. Birdwood, C.S.I., C.I.E., A.D.C., G.O.C., Kohat Brigade, reported :—" General efficiency thoroughly satisfactory. The Battalion is quite fit to take the field at any time. It is well drilled and officered, and has confidence in itself." The General Officer Commanding Northern Command added :—" Not seen this year, but I know this battalion to be in the very high state of efficiency reported by the Brigadier-General, which does much credit to Lieutenant-Colonel Campbell, and those serving under him."
Kohat	... 1911 ...	Brigadier-General Birdwood, C.S.I., C.I.E., A.D.C., G.O.C., Kohat Brigade, reported :—" Musketry . . . signalling . . . I doubt if the general efficiency attained in both the above subjects can be beaten in the Indian Army. " The men are exceptionally well set up, due a great deal to the great trouble which is taken with the recruits. All ranks take much pride in the smart turn-out of the Regiment, and the men of the 26th are always an example to the other corps in this respect. . . . " I cannot speak too highly of the general efficiency of the 26th Punjabis, of which, perhaps, as good a proof as any is the extraordinary respect which all other corps in the garrison have for it. A really fine spirit pervades the whole Regiment, due to the fact that the Regiment knows its capabilities, the result of every detail receiving attention, and training being carried out thoroughly and with spirit."

APPENDIX II—(*continued*)

PLACE.	DATE.	NAME OF INSPECTING OFFICER AND REMARKS.
Hangu, Thall and Fort Lockhart ...	April, 1911...	Lieutenant-General Sir Jas. Willcocks, K.C.B., Commanding Northern Army, said :—" A really good battalion. I saw them on a hill warfare day at Saraghari this winter, and at field firing. They are a cheery, well set-up, well-disciplined corps. The recruits are far and away the best I saw in the Northern Army this year. Fit for active service." The Commander-in-Chief's remarks were :—" An excellent report."
Kohat Brigade	Jan., 1912...	Brigadier-General W. Birdwood, C.B., C.S.I., C.I.E., A.D.C., Commanding Kohat Brigade, reported :—" I feel I can add or detract nothing from the excellent report I made on the 26th Punjabis last year. The greatest attention is paid to all details both in training and in all that affects the well-being and moral of the men. The result is a very fine Regiment in every respect."
Kowloon, China	9/4/13 ...	Major-General Anderson, C.B., Commanding South China, inspected the Regiment, and gave an excellent report.
Kowloon ...	1914 ...	Major-General Kelly, C.B., Commanding South China, gave an excellent report, on which the General Officer Commanding Southern Command, and Commander-in-Chief in India commented respectively, as follows :— " A good report." " Very satisfactory."
Kowloon ...	5-6/2/15 ...	Major-General Kelly, C.B., Commanding South China, reported :—" All that can be desired. Regiment quite fit, and longing for service in the field. I hope after a month's leave in India they will be sent to France. There is no better fighting unit in the Indian Army."
	1916 - 1919	No reports owing to Great War.

Peshawar	... 29/3/20 ...	Brigadier-General C. C. Luard, C.B., C.M.G., Commanding 1st Infantry Brigade, Peshawar, reported as follows :—" This unit has only lately returned from field service in Mesopotamia, and owing to the whole Battalion having been on leave, training is in a backward state. . . . The material is good, men' generally are of good physique and well set-up. . . . There is *esprit de corps* and a good spirit in the Battalion ; and all ranks seem keen to do their best, and given an opportunity, I am confident that they will soon be fit for service. . . ."
		Major-General Sir G. de S. Barrow, K.C.B., K.C.M.G., Commanding 1st (Peshawar) Division, added :—" A good set of men in the ranks, and some very good Indian officers. . . .
		" The Battalion is energetically commanded, and a good tone pervades all ranks."
Peshawar	... 10/2/21 ...	Brigadier-General C. C. Luard, C.B., C.M.G., Commanding 3rd Infantry Brigade, reported as follows :—" Training satisfactory ; fit for service."
		Major-General Sir G. de S. Barrow, K.C.B., K.C.M.G., Commanding 1st (Peshawar) Division, reported :—" A good battalion, well commanded, well turned out, and well drilled. Organization, *esprit de corps*, and training are all satisfactory."
		General Sir W. Birdwood, K.C.B., K.C.S.I., C.I.E., General Officer in Command, Northern Command, added :—" A battalion which as of old keeps up the great reputation it has held for many years in the Indian Army."
Waziristan	... 12/5/22 ...	The General Officer Commanding Wazir Force, on 12/5/22 inspected a company (" D " Company, Dogras, under Captain Steer and Lieutenant Young) at drill on the parade ground at Sorarogha, and expressed great satisfaction, saying it was the best company he had seen in the Force.
	18/5/22 ...	The Commandant, Vickers' Gun School, thoroughly tested No. 20 Platoon (Vickers' guns, under Lieutenant Bwye and Havildar Nand Singh), and stated that it was the best trained platoon he had inspected.

APPENDIX II—(*continued*)

PLACE.	DATE.	NAME OF INSPECTING OFFICER AND REMARKS.
Waziristan (*contd.*)	17/4/23 ...	The General Officer Commanding Wazir Force, Major-General Sir Torquihil Matheson, K.C.B., on the occasion of the Regiment leaving the force, wrote to the Commanding Officer (Lieutenant-Colonel P. S. Stoney) as follows :

" Before your Battalion leaves this Force I wish to say good-bye to you and to thank you and all ranks under your command for the good work you have done in the force. Since your arrival in Waziristan in January, 1921, you have served on the Lines of Communication at Kotkai and Sorarogha until March, 1923, when you formed part of the column which marched from Jandola to Wana to withdraw the S.W. Scouts and instal them at Sarwakai.

" Throughout this period you have been often in action against the enemy, and by virtue of your high courage, military efficiency, and discipline have earned both the respect of the enemy and the admiration of your comrades.

" I also wish to congratulate you on your smartness on parade ; always a sign of *esprit de corps* and of a good regiment.

" I regret to see that you suffered 71 casualties while in Waziristan, of which no less than 21 were killed in action.

" Now that you are proceeding to enjoy (I hope) a well-earned rest in a peace station, I wish you very good luck and feel sure you will always maintain and add to the fine record you have won here and elsewhere."

APPENDIX III.

MUSKETRY SUCCESSES.

DATE.	MATCH, ETC.	RESULT.
1888	B.P.R.A....	23 men obtained prizes—Lance-Naik Bela Singh winning 2nd prize (Rs100) in match 34.
1888	Annual Course ...	Improvement of figure of merit from 110.30 with Snider to 145.75 with M.H.R.
1889	B.P.R.A.—Inter - Regimental	First—Cup value Rs350. 46 men obtained prizes, and teams won matches I, II, IV, V.
1890	B.P.R.A.	Naik Bela Singh won the Championship.
1891	C.-in-Cs. Prize ...	First.
	B.P.R.A.—Inter-mental Match Regimental	First.
		Naik Makhmad won the Magdala gold medal.
		Telegraphic congratulations were received from His Excellency the Commander-in-Chief and the General Officer Commanding Peshawar.
		The Regiment won in all Rs672, including the Inter-Regimental and N.C.Os. match.
	Central Meeting, Poona	Kolharpur Cup—First.
		Naik Bela Singh, silver medal in aggregate.
	Central Meeting, Meerut	Havildar Shah Khan won the Prince of Wales' badge and prize of Rs75 and the Viceroy's silver medal ; also lost on the tie for the silver medal (aggregate).
		Naik Bela Singh was 4th in the latter.
1892	Annual Course ...	Figure of merit in (2nd period collective) 56.57. The Regiment was 11th in the N.A.
	B.P.R.A.—Inter-Regtl....	First.
	N.C.Os. ...	First.
	Volley firing ...	Third.
	Inter-Company ...	Fourth and Thirteenth.
		The Regiment won a total of Rs572 and had a 2nd and 4th in the silver medal.

APPENDIX III—(continued)

DATE.	MATCH, ETC.	RESULT.
1892 ...	Central Meeting, Meerut Inter-Regimental	Third.
1892-93...	C.-in-Cs. Match	First score 858.
	Annual Course	Figure of merit 51.67. Fever was cause of decrease.
	B.P.R.A., Local... ...	Total prizes, Rs589.
	Central Meeting—Inter-Regimental	First—Cup value Rs600. Championship—Lance-Havildar Makhmad.
30/3/94...	C.-in-Cs. Match	First—Score, 826.
	" Honour and Glory " Match	First—Score, 921. Jemadar Magar Singh, 98 points.
	B.P.R.A. Local Meeting, Inter-Regimental Match	First—Also Rs475 in various prizes.
	Central Meeting—Inter-Regimental Match ...	First—Silver cup value Rs600. Championship—Jemadar Magar Singh.
1895 ...	C.-in-Cs. Match	First—Score, 845.
1896-97 ...	B.P.R.A....	No men competed as most were on leave and furlough after Egypt.
Feb.-Mar.,1898	B.P.R.A.—Central Meeting, Meerut	Won a handsome Silver Cup presented by Sir W. Lockhart ; also large share of individual prizes.
Dec., 1898 ...	B.P.R.A....	First six places in Championship, but no cups. Some of the best regiments were not represented this year owing to service on North-West Frontier.
1898-99 ...	Annual Course	Figure of merit, 64.
Dec., 1899 ...	B.P.R.A.—C.-in-Cs.	Second (2 marks behind 29th P.I.)
	Meerut Cup	First—Sir Power Palmer's Cup, value Rs300.
1900 ...	B.P.R.A.—L.G.Cs. Cup	First.
1901 ...	Annual Course	Percentage 62.
	B.P.R.A.—Meerut ...	Subadar Major Magar Singh won the Native Army Championship and men of the Regiment won over Rs1,200 in prizes, including 1st, 2nd, 3rd, 4th, 5th and 6th in long range aggregate, and 1st, 2nd, 3rd, 5th, 6th and 7th in the Championship.

1901	...	Cawnpore Cup, N.C.Os.	First.
Dec., 1902	...	Annual Course	Percentage 66.
		Peshawar Local Meeting	Officers' Match—First.
Dec. 1903	...	Annual Course	Percentage 62.
1904	...	Annual Course	Percentage 63.
1905	...	B.P.R.A.—Meerut Cup...	First.
		C.-in-Cs. Cup	" G " Company were 2nd, and " H " Company 3rd.
Mar., 1906	...	Annual Course	Percentage 65.
		B.P.R.A....	Extended Order Competition—First.
			No other large prizes as the Regiment was bad with fever.
		Annual Course	Percentage 67.
Mar., 1907	...	Annual Course	Percentage 67.
		C.-in-Cs. Cup Coy. team	First equal (" G " Company), but lost on firing it off.
		Cawnpore Cup, " D " Coy. team	First. 4th Double Company (" G " and " H ")—Challenge Cup, and also a cup to keep.
		B.P.R.A....	2 British Officers, 23 N.Os. and N.C.Os. and men competed, and all but one or two made their expenses.
1908	...	C.-in-Cs. Cup	Second team from " G " Company (equal with 36th Sikhs).
1910	...	C.-in-Cs. Cup	Second—Team of men from different companies.
		Cawnpore Woollen Mills Cup	Second—Team from " G " Company (Afridis).
24/5/10	...	Empire Cup Match ...	Tenth—With prizes of £15. Chance team of 2 N.Os., 188 N.C.Os. and men.
1911	...	Magdala Gold Medal ...	No. 4829 Havildar Allah Din.
21/5/12	...	Empire Cup Match ...	Fourteenth—With prize of £10.
1/5/13	...	Empire Cup Match ...	Seventh—With prize of £15.
18/5/13	...	Hatton Cup Competition, Hong-Kong	First—With score of 134 points, which was more than double the score of other units competing.
8/4/14	...	Hatton Cup Competition, Hong-Kong	First again with 181 points. The next unit had 117 points. Team of 40 N.C.Os. and men under Subadar Jan Gul.
May, 1914	...	Empire Cup Match ...	Fifth—Score, 2,480.

Since 1914 the Regiment has had no opportunity of entering for Musketry matches.

APPENDIX IV.

SIGNALLING RESULTS.

PLACE AND DATE.	POSITION IN INDIAN ARMY AND FIGURE OF MERIT.	REMARKS.
Peshawar, Mar, 1898	Fourth—468·29	The results of the inspection show that the men have been carefully trained. The work throughout is very accurate, and the rates good. I consider all the signallers to be thoroughly efficient.
Jullundur Feb., 1899	Tenth—489·49	The results of the inspection are very satisfactory ; great interest is evidently taken in signalling in the Regiment, as is evidenced by the large amount of private equipment possessed. The flag drill was excellent. The sending by the selected signallers on the helio and the lamp was very good and the reading in the helio pair test very accurate.
Jullundur Feb., 1900	Eighty-fifth—326·26	Below satisfactory standard. "Signalling instructions" not corrected to date. Messages inaccurate. A new system was introduced, and the Signalling Officer was on furlough.
Malakand Dec., 1900	First—1090·00	The inspection was a most satisfactory one. The men are exceedingly well up in book work and were very smart in carrying out each test. This high state of efficiency reflects great credit on the Instructor (Captain Lawson). The selected signaller is an exceptionally good sender (Havildar Mul Singh).
Malakand Oct., 1901	First—973·00	I find it difficult to accord sufficient praise for the admirable manner in which the signallers of this Regiment have been trained. I consider that they have reached a higher degree of efficiency than ever was expected from the Native Army, and their work may be put on a par with that of a very good British regiment. The Regiment could, without difficulty, supply two special

signalling units for field service, and the results of the inspection reflect the greatest credit on all concerned.

The Regiment may be quoted as an ideal pattern to the remainder of the Native Army, as regards general signalling efficiency.

Peshawar ... Jan., 1903	Second—949·98	...	The Regiment continues to maintain its reputation for possessing well trained and intelligent signallers. Flag drill was smartly carried out and the theoretical knowledge displayed by the men was excellent and above the average of a British regiment. The manipulation of all instruments was faultless and the discipline at work very good. Eighteen signallers were sent to Seistan with the Boundary Delimitation party this year.
Miranshah ... Mar., 1904	Second—1003·00	...	The very high figure of merit obtained by the signallers testifies to the accuracy of their reading and sending on all instruments. The work at inspection was in addition smoothly performed and was faultless throughout. The general knowledge of the men is excellent and they are thoroughly proficient in all duties of signallers.
D. I. Khan... Mar., 1905	Second—766·00	...	The results obtained are most satisfactory and show the highest state of efficiency all round. The whole of the work was carried out with accuracy and quickness, and the discipline at work was perfect.
D. I. Khan ... Nov., 1906	Third—793·92 ...	...	A highly trained signalling unit which fully maintains its high standard. The distinguished standard has been attained ; but owing to the failure in reading of one of the signallers, the Regiment cannot be classified as such. It is, however, specially commended for its excellent performance.
D. I. Khan... Oct., 1907	No position assigned or figure of merit given. System under alteration.		A very efficient and smart unit, and appear to be very well trained. The Supernumeraries are very smart and really good all round.

K

APPENDIX IV—*(continued)*

PLACE AND DATE.	POSITION IN INDIAN ARMY AND FIGURE OF MERIT.		REMARKS.
Kohat, 1908	No place assigned	...	The work was very good throughout. . . . The men are a very well trained body of signallers, who will acquit themselves well wherever they are employed.
Kohat, 1909	No place assigned	...	The signallers have been trained to a very high standard of excellence individually, and have fully maintained the fine reputation of the Regiment in previous years in this respect. . . . Havildar Buta Singh, the senior assistant instructor, deserves special credit for his individual exertions in the training of this splendid body of signallers, who have but few equals in the Indian Army.
Kohat, 1910	" Excellent " ...	...	The signallers of the 26th Punjabis, all of whom qualified as 1st class, are a splendidly trained body of men, who have few equals in the Indian Army.
Kohat, 1911	—		The results of the inspection are excellent. The signallers have thoroughly upheld the high reputation for all round excellence in the technical training they have enjoyed for so many years past.
Hong-Kong, 1912 ...			A very excellent report.
1914 ...			A very good inspection report in which all (24) signallers qualified as 1st class.
1914 to 1923.			During the war regular inspections of signallers were not made. Our signallers continued to display great efficiency in practical field work, and on their first inspection in 1923, after coming from Waziristan, they scored a figure of merit of 97·20, the second highest in the district, and also were reported " very satisfactory " in field work, being the only unit in the district to attain this high commendation. These results reflected great credit on Subadar-Major Buta Singh and Havildar Natha Singh (S.A.I.).

APPENDIX V.

Subadar-Majors.

NAMES.	REMARKS.	DATES.		
		ENLISTMENT.	APPT. OF SUB.-MAJ.	PENSIONED.
Abdullah Khan, Sardar Bahadur, Malik Din Khel Afridi.	Native Adjutant, 1857. Order of Merit, 3rd Class. 1st Class Order of British India. Over 40 years' service, 33 of which in the 26th Punjabis. Pension 1890. Died 1907.	15/6/57 ...	1858 ...	6/1/90
Lakhmir Singh, Sardar Bahadur, Sikh	Enlisted in 6th Punjab Infantry 14/5/52. Transferred as Drill Instructor. 1st Class Order of British India. Pension 1893. Died 1904. Over 41 years' service, 36 in 26th Punjabis.	15/6/57 ...	1890 ...	1/10/93
Mansur Khan, Sardar Bahadur, Yusufzai Pathan.	3rd Class Order of Merit, 1886. 1st Class Order of British India. 36 years' service. Died 13/10/94.	16/1/58 ...	2/10/93 ...	1894
Jiwan Singh, Bahadur, Sikh	33 years' service. Pension 1897. Still living.	7/4/63 ...	14/10/94 ...	1/4/97
Sher Baz, Bahadur, Malik Din Khel Afridi.	2nd Class Order of British India. Pension 1901. Still living.	20/10/69 ...	2/4/97 ...	20/10/01
Magar Singh, Sardar Bahadur, Sikh	2nd Class Order of British India, 1903. Orderly Officer to H.M. the King-Emperor, 1904. 1st Class Order of British India, 1907. Pension 1906. Died 1911.	2/6/73 ...	21/10/01 ...	11/11/06

K*

APPENDIX V—(*Continued*)

NAMES.	REMARKS.	DATES.		
		ENLISTMENT.	APPT. OF SUB.-MAJ.	PENSIONED.
MAKHMAD, Malik Din Khel Afridi.	Winner of many shooting prizes.	15/6/85 ...	12/11/06 ...	5/1/10
MUHAMMAD AKBAR, Bahadur, Afridi.	2nd Class Order of British India.	27/1/87 ...	6/1/10 ...	5/1/16
HARNAM SINGH, Sikh.	Invalided from F.S. in Mesopotamia on 9/3/16, but rejoined.	17/6/87 ...	6/1/16 ...	1918
ISHAR SINGH, Sikh.	Extra S.M. on F.S. *vice* Harnam, S., invalided, 1917 and 1918, served at Depot. Pensioned as Honorary Lieutenant.	5/5/88 ...	9/3/16 ...	11/10/20
MAWAZ KHAN, I.D.S.M., Punjabi Musalman	Acting S.M. on F.S. *vice* Ishar Singh, 1917 and 1918. Became S.M. when Ishar Singh retired, 1920. Received I.D.S.M. for service in Mesopotamia.	19/1/01 ...	11/10/20 ...	May,1922
BUTA SINGH, Sikh	Present Subadar-Major.	20/3/96 ...	May 1922	

APPENDIX VI.

COPIES OF INDIAN ARMY LISTS, 1858, 1907, 1922, AND FINAL POSTING, 1923.

1858.

Corrected up to 10/7/58 (*only issue in* 1858).

18th Infantry.—Raised at Peshawar. At Campbellpore.

NAMES, RANK AND CORPS.	ARMY RANK.	WHEN APPOINTED.	REMARKS.
Williamson, Lieut. J., 49th N.I. (Comdt., 9th Punj. Infy.) ...	13/4/48 ...	31/12/57 ...	Offg. Comdt.
King, Capt. H., 39th N.I. ...	10/12/57 ...	25/5/58 ...	
Roberts, Lieut. T. L., H.M. 87th Fus.	27/12/50 ...	31/12/57 ...	
Hunter, Lieut. C., H.M. 81st Foot	1/8/51 ...	30/6/58 ...	
Van der Gucht, Lieut. T. E., 5th N.I.	9/8/54 ...	31/12/57 ...	
Stewart, Asst.-Surg. J. L., M.D....	4/8/55 ...	30/6/58 ...	

June, 1907.

Regt. Centre—Lahore. *Linked with*—20th and 21st Punjabis.

Uniform—Drab. *Facings*—Scarlet.

26th Punjabis.

Raised at Peshawar in 1857, by Capt. H. T. Bartlett, as the 18th Regt. of Punjab Infy. Became the 30th Regt. of Bengal Native Infantry, 1861 ; the 26th Regt. of B. N. I., 1861 ; the 26th (Punjab) Regt. of B. N. I., 1864 ; the 26th (Punjab) Regt. of Bengal Infy., 1885 ; the 26th Punjab Infy., 1901 ; received present designation, 1903.

" Afghanistan, 1878–79 " ; " Burma, 1885–87."

Dera Ismail Khan.—Arrived 25/10/04 from Miran Shah (under orders to Kohat).

Composition—4 Companies of Sikhs, 2 of Afridis, 2 of Punjabi Musalmans.

Colonel.—Major-General Lewis Dening, C.B., D.S.O., 13/5/04.

APPENDIX VI—(continued)

FIRST COMMN.	NAMES AND RANK.	ARMY RANK.	PRESENT APPT. IN REGT.	REMARKS.
		Commandant.		
10/5/82 ...	Campbell, Lt.-Col. A. A. E., *p.*, Q.	29/10/06 ...	29/10/06 ...	Lv. ex I., 8 mos., 13/3/07.
		Double Company Commanders.		
22/10/81 ...	Bernard, Maj. E. H., *p.*	10/7/01 ...	12/7/06 ...	Second-in-Command.
25/8/86 ...	Walton, Maj. L. B., *p.*	25/8/04 ...	1/5/00 ...	
11/2/88 ...	Rattray, Maj. C., *p.s.c.*, Q. ...	11/2/00 ...	1/5/00 ...	Offg. Bde.-Maj., Derajat Bde.
10/11/88 ...	Thompson, Maj. I. F. R., *p.* ...	10/11/06 ...	1/5/00 ...	Circle Registn. Officer.
8/5/90 ...	Lawson, Capt. O. H., *p.*	8/5/01 ...	23/11/00 ...	
		Double Company Officers.		
1/9/95 ...	Little, Capt. I. M.	1/3/06 ...	13/2/03 ...	
22/1/98 ...	Turnbull, Capt. G. O.	22/1/07 ...	1/7/96 ...	Adjt., 1/4/04.
27/7/98 ...	Grey, Lieut. A. J. H.	27/10/00 ...	1/4/00 ...	Pol. Dep. (probation).
25/1/99 ...	Stoney, Lieut. P. S.	25/4/01 ...	22/4/00 ...	Qr.Mr., 23/3/06 Lv. ex I., to 30/11/07.
7/3/00 ...	Ivens, Lieut. H. T. C.	10/7/01 ...	16/1/02 ...	A.R.S.O. for Pathans.
11/8/00 ...	Cargill, Lieut. R. J.	23/3/02 ...	24/8/03 ...	
14/9/01 ...	Salusbury, Lieut. R. T. G. ...	14/12/03 ...	5/12/03 ...	
8/2/02 ...	Maude, Lieut. E. A.	8/5/04 ...	17/1/04 ...	Lv. ex I., m.c., 8 mos., 20/2/07.
19/8/03 ...	Bennett, Lieut. O. D.	19/11/05 ...	1/9/06 ...	
9/1/04 ...	Anderson, Lieut. G. W.	9/4/06 ...	13/10/06 ...	
		Medical Officer.		
26/7/02 ...	O'Keeffe, Capt. D. S. A., M.B. ...	26/7/05 ...	8/8/06 ...	
		Subadar-Major.		
		Jemadar.	Subadar.	Subadar-Major.
15/6/85 ...	Makhmad (56)	26/6/97 ...	27/3/03 ...	11/11/06

Subadars.

13/3/01	...	Yar Muhammad Khan (72a)	...	13/3/01	...	13/3/03	...	
15/7/00	...	Lachman Singh (72a)	...	15/7/00	...	25/6/03	...	
27/1/87	...	Muhammad Akbar (56, 72a, 72b)		17/12/99	...	16/10/03	...	
17/6/87	...	Harnam Singh	...	25/6/03	...	2/12/04	...	
21/10/89	...	Ali Haidar ...	...	15/11/02	...	17/1/06	...	(Supy.) with Somaliland Rifles.
5/5/88	...	Ishar Singh (58, 68)	...	16/3/04	...	11/11/06	...	

Jemadars.

16/7/91	...	Jan Gul (56)	...	27/3/03	...	Native Adjt. 24/7/03.
19/6/87	...	Waryam Singh	...	1/4/03		
16/12/90	...	Basimullah	...	25/6/03	...	
5/2/91	...	Tura Baz ...	...	12/3/04	...	...
26/3/04	...	Muhammad Quresh (56)	...	26/3/04	...	...
23/11/87	...	Mal Singh (56, 72a)	...	2/12/04	...	...
4/9/05	...	Harnam Singh	...	17/1/06	...	On probation.
23/7/87	...	Sham Singh (56, 58)	...	11/11/06		...

April, 1922.

FIRST COMMN.	NAMES AND RANK.	ARMY RANK.		REMARKS.

Commandant.

| 25/1/99 | ... | Stoney, Lt.-Col. P. S., *p.s.c.* | ... | 8/2/22 | ... | ... | |

Second-in-Command.

Company Commanders.

14/9/01	...	Salusbury, Maj. R. T. G. (*s.c.*)	...	1/9/15	...	—	Lv. ex I., to 22/1/23, pending retirement.
8/2/02	...	Maude, Maj. E. A., D.S.O.	...	1/9/15	...	—	Second-in-Command (tempy.)
7/5/02	...	Edwards, Maj. E. ...	...	7/5/17	...	—	Lv. ex I., 1 yr., 11/2/22.
19/8/03	...	Bennett, Maj. O. D.	...	19/8/19	...	—	With S. Waziristan Scouts.

APPENDIX VI—(*continued*)

Company Officers (8).

FIRST COMMN.	NAMES AND RANK.	ARMY RANK.		REMARKS.
9/1/04	Anderson, Maj. G. W.	9/1/19	—	With Ind. State Forces.
9/9/08	Fulton, Capt. J. D. M.C.	1/9/15	—	Lv. ex I., m.c., to 8/6/22.
3/9/10	Shearer, Capt. F. E., M.C.	1/9/15	—	
1/10/14	Farwell, Capt. G. A. L., M.C.	1/10/18	—	Adjt.
29/6/16	Conner, Capt. T. S.	1/4/19	—	Attd.
10/9/15	Pollock, Capt. A. H., M.C.	10/9/19	—	Attd.
24/12/14	Geary, Capt. H. V., M.C.	24/9/19	—	Attd.
20/10/18	Steer, Capt. J. M. B.	20/10/19	—	Attd.
13/1/16	Maclaren, Capt. W. de B.	13/1/20	—	Attd.
30/5/16	Green, Capt. W. G.	30/5/20	—	Attd.
29/1/16	Thomas, Capt. J. E.	29/10/20	—	Attd.
14/11/16	Williams, Capt. B. F.	14/11/20	—	Attd.
30/1/17	Robertson, Capt. H. L. C.	30/1/21	—	
25/7/17	St. Ruth, Capt. G. H. B. M.C.	25/7/21	—	Attd.
21/12/17	Rootes, Lieut. A. H.	21/12/18	—	Attd.
27/3/18	Young, Lieut. J. W. F.	27/3/19	—	Attd.
27/3/18	Hughes, Lieut. T. L.	27/3/19	—	Attd., Lv. ex I., 8 m., 6/3/22.
81/8/18	Lane, Lieut. C. M.	31/8/19	—	Attd.
16/12/18	Whitehead, Lieut. A. W. N.	16/12/19	—	Attd., Lv. ex I., to 27/3/22.
15/6/18	Gray, Lieut. G. E.	15/3/20	—	Attd.
15/4/19	Wiseman, Lieut. D. J. C.	15/4/20	—	Attd.
29/1/20	Fitzpatrick, Lieut. B. P. S.	29/1/21	—	Attd.
29/1/20	Vincent, Lieut. M. V.	29/1/21	—	Attd.
29/1/20	Bwye, Lieut. M.	29/1/21	—	Attd.
22/7/18	Martin, 2/Lieut. J. E. L.	22/7/18	—	Attd.

Attached.

18/9/15 ...	Driver, Capt. R. A.	18/9/19 ...	—	2/19th Punjabis.
17/7/16 ...	Edwards, Capt. M. F.	17/7/20 ...	—	1/19th Punjabis.

Subadar-Major.

		Jemadar.	Subadar.	Subadar-Major.
19/1/01 ...	Mawaz Khan, I.D.S.M. (56) ...	1/1/12 ...	1/5/15 ...	11/10/20.

Subadars.

		Jemadar.	Subadar.	
20/3/96 ...	Buta Singh (56, 58, 72a)	21/10/10 ...	9/3/16 ...	
6/6/01 ...	Suraj Din (51, 58)	1/2/13 ...	5/4/16 ...	
20/8/03 ...	Dasondha Singh	12/2/16 ...	1/7/17 ...	
18/4/03 ...	Sardar Khan, I.D.S.M.	1/10/16 ...	17/12/17 ...	
22/12/96 ...	Sundar (72a)	30/12/15 ...	29/10/18 ...	
1/11/99 ...	Nag (66a)	1/11/15 ...	15/1/20 ...	
14/7/02 ...	Ghamandhi Singh	8/12/19 ...	8/9/20 ...	
24/4/03 ...	Hayat Muhammad, I.O.M. (48a, 51)	2/1/17 ...	11/10/20 ...	

Jemadars.

		Jemadar.	Subadar.	
20/8/03 ...	Narayan Singh, I	1/11/17 ...	—	Ind. Adjt.
9/2/99 ...	Narayan Singh, II	2/5/18 ...	—	
19/7/02 ...	Duni Chand	1/1/20 ...	—	
28/5/07 ...	Udmi	9/5/20 ...	—	
22/5/01 ...	Gurditta	1/8/20 ...	—	
22/7/11 ...	Motla	31/8/20 ...	—	
3/4/08 ...	Bhalu Singh	1/9/20 ...	—	
8/2/01 ...	Sharam Singh	15/10/20 ...	—	Ind. Qr.Mr.
18/5/03 ...	Bhagat Singh	1/11/20 ...	—	
21/4/06 ...	Nadar Ali	1/11/20 ...	—	
11/9/07 ...	Jaswant Singh	1/4/21 ...	—	

APPENDIX VI—*(continued)*.

2ND BN. 15TH PUNJAB REGIMENT.
Final Posting of Officers, 1923.

Lt.-Col. P. S. Stoney ...	Commandant	—
Maj. H. T. C. Ivens ...	Second-in-Command ...	Seconded
Maj. E. Edwards ...	Coy. Commander ...	—
Maj. O. D. Bennett ...	Coy. Commander ...	—
Maj. G. W. Anderson ...	Coy. Commander ...	Seconded
Capt. J. D. Fulton, M.C.	Coy. Commander ...	—
Capt. F. J. H. Nugent, D.S.O.	Coy. Commander ...	Seconded
Capt. J. E. Shearer, M.C.	—	Seconded
Capt. G. A. L. Farwell, M.C.	—	Seconded
Capt. T. S. Conner ...	—	—
Capt. A. H. Pollock, M.C.	—	—
Capt. J. M. B. Steer ...	—	Seconded
Capt. H. L. C. Robertson	—	—
Capt. A. H. Rootes ...	—	—
Capt. C. M. Lane ...	—	Seconded
Lt. R. Cook	—	Seconded
Lt. D. J. C. Wiseman ...	—	—
Lt. J. E. L. Martin ...	—	—
Lt. W. J. Blake	—	—
Lt. R. F. Raikes ...	—	Seconded
Lt. B. P. S. Fitzpatrick		—

Attached.

Capt. A. G. I. A. Goddard.
Lt. C. G. Wilson.
Lt. M. Bwye.

APPENDIX VII.

Roll of Honours and Awards during the Great War,
191418.

Brevet-Colonel.

Lt.-Col. L. B. Walton.

Distinguished Service Order.

Major G. O. Turnbull.
Major E. A. Maude.

Military Cross.

Capt. J. D. Fulton.
Capt. J. E. Shearer.
Capt. G. A. L. Farwell.

Order of British India, 2nd Class.

Subadar Feroz Khan, 46th P.

Indan Order of Merit.

Subadar Jan Gul.
759, Naik Gulab Khan.
Jemadar Hayat Muhammad.
Subadar Sher Akhmad.

Indian Distinguished Service Medal.

1199 Lance-Naik Channan Singh.
1144 Lance-Naik Ghulam Haidar.
845 Lance-Naik Bawa Singh, 21st P.
1755 Sepoy Hazara Singh.
Jemadar Ghulam Husain.
Subadar Sardar Khan.
669 Sepoy Waryam Singh.
Acting Subadar-Major Mawaz Khan.
Jemadar Mir Dast.

Indian Meritorious Service Medal.

Jemadar Narain Singh (III).
415 Naik Harnam Singh.
581 Naik Dial Singh.
1165 Naik Moghul Khan.
743 Lance-Naik Ran Singh.
4617 Lance-Naik Labh Singh.

APPENDIX VII—*(continued)*.

Mentions in Despatches.

Lt.-Col. (tempy. Brig.-Gen.) L. B. Walton.
Major E. A. Maude (twice).
Capt. J. D. Fulton (six times).
Capt. J. E. Shearer.
Lieut. G. A. L. Farwell.
Lt.-Col. A. D. Cox.
Capt. G. W. Anderson (twice).
Lieut. D. S. Warren.
Jemadar Hayat Muhammad.
Subadar Feroz Khan, 46th P.
Jemadar Ghulam Husain (twice).
Subadar Sardar Khan.
Subadar Dasaundha Singh.
Acting Subadar-Major Mawaz Khan.
Major G. O. Turnbull.
Major P. S. Stoney.
4860 Havildar Kesar Singh.
4710 Havildar Sharm Singh.
415 Naik Harnam Singh (twice).
402 Naik Kartar Singh.
2385 Naik Khan Beg, 46th P.
1199 Lance-Naik Channan Singh.
1144 Lance-Naik Ghulam Haidar.
743 Lance-Naik Ran Singh.
2572 Sepoy Dost Muhammad.
1755 Sepoy Hazara Singh.
965 Sepoy Mul Singh.
1118 Havildar Jawand Singh.
926 Havildar Kirpa Singh.

Croix de Guerre.

Capt. J. E. Shearer.

Persian Order of the Lion and Sun, 3rd Class.

Capt. J. D. Fulton.

Serbian Decoration.

4711 Sepoy Sher Singh.

APPENDIX VIII.

Table of Casualties and Rolls of Casualties to British Officers and Indian Officers during the Great War, 1914–18.

Casualties with the Regiment.

Killed—died of wounds, missing believed dead, 55.
Wounded, 229.
Died of disease, 36.

British Officers.

Lt.-Col. L. B. Walton : Died as result of service in Mesopotamia, May, 1917.
Lt.-Col. I. F. R. Thompson : Killed in action, 26/1/17.
Major O. H. Lawson : Died of wounds, 17/3/16.
Major I. M. Little : Wounded, 8/3/16.
Major G. O. Turnbull (attd. Royal Scots) : Wounded in France, 1915.
Major H. T. C. Ivens : Wounded, 26/1/17.
Capt. E. A. Maude : Wounded, 8/3/16.
Capt. R. J. Cargill : Wounded (shell-shock) and invalided as result of service in Mesopotamia.
Capt. J. E. Shearer : Wounded, 24/2/17.
Lieut. J. H. W. Stevenson (46th P.) : Died of wounds, 5/2/17.
Capt. C. C. T. Teape : Died at Peshawar as result of service in Mesopotamia, March, 1920.
Capt. R. T. G. Salusbury (attd. 120th Infy.) : Wounded, Shaiba, April, 1915.
Lieut. C. J. Torrie (attd. Royal Scots) : Wounded, France, 1915.
Lieut. Balkrishna (I.M.S.) : Wounded, 26/1/17.
Lieut. H. D. Drysdale (attd. 2nd Royal Scots) : Killed in France, 31/8/15.
Lieut. H. C. W. Dillon (attd. 24th P.) : Killed in Mesopotamia, 1915.

Indian Officers.

Subadar-Major Harnam Singh : Wounded, 8/3/16.
Subadar Mal Singh : Wounded, 8/3/16.
Subadar Harnam Singh : Wounded, 8/3/16.
Jemadar Shamir Singh : Killed in action, 26/1/17.
Jemadar Harnam Singh : Killed in action, 26/1/17.
Jemadar Sher Baz (21st P.) : Wounded, 8/2/17.
Jemadar Tika Khan : Died of disease, February, 1917.
Subadar Amar Singh (21st P.) : Wounded, 26/1/17.
Jemadar Suraj Din : Wounded, 26/1/17.
Jemadar Sham Singh : Wounded, 26/1/17.

APPENDIX IX.

HONOURS AND REWARDS AWARDED TO THE MEN OF 26TH
PUNJABIS IN WAZIRISTAN, 1921.

Indian Distinguished Service Medal.

Awarded for gallantry in action :—
    3249 L./Naik Des Ram, " B," 16/12/21.
    2279 Naik Labha, " D," 12/1/22.
    4442 Sepoy Kunji, " B," 16/1/22.

*Mentioned in Despatches of H.E. Lord Rawlinson of Trent,
   C.-in-C. in India, dated 20/7/23.   (Gazette of India, No. 995).*

Lieut.-Col. P. S. Stoney.
Major G. W. Anderson.
Capt. G. A. L. Farwell, M.C.
Major (A./Lieut.-Col.) E. A. Maude, D.S.O.
Lieut. A. H. Rootes.
Subadar-Major Buta Singh.
1013 Havildar Balwant Singh.
Jemadar Bhagat Singh.
930 Havildar Harnam Singh.
112 Havildar Nand Singh.
1253 Naik Karm Ilahi.
926 Havildar Kirpa Singh.
2012 Coy. Havildar Major Lehru.
3804 Sepoy Samwalia.

*I.Os., N.C.Os., and Men of the 26th Punjabis awarded
   " Wazirforce" Commander's Certificate for " Devotion to
   Duty " in 1922.*

Jemadar Bhagat Singh.
1013 Havildar Balwant Singh.
 930 Havildar Harnam Singh.
1910 Sepoy Teja Singh.
4707 Havildar Mohna Ram.
4880 Havildar Aman Singh.
3804 Sepoy Samwalia.
4890 Sepoy Bhawani.
1116 Havildar Mohd Sher.
1253 Naik Karam Ilahi.
2012 Havildar Lehru.
2347 Havildar Suram Singh.
1295 Havildar Abbas Khan.
 926 Havildar Kirpa Singh.

142

515 Havildar Matab Singh.
809 Havildar Nand Singh.
913 Naik Sunder Singh.
1685 L./Naik Nek Mohammed.
2717 L./Naik Nathu.
2716 Sepoy Mirza Khan.
2368 Sepoy Wali Mohammed.
3461 Sepoy Shiama.
3877 Sepoy Kapur Singh.
2574 Sepoy Khan Mohammed.

*Men awarded Waziristan Force Commander's Certificate for the
     Year 1922–23.*

" A " Company.

467 Sepoy Puran Singh.
1777 Sepoy Bachan Singh.

" B " Company.

2710 Havildar Shiv Lall.
 296 Havildar Basti.
2265 Havildar Ramphal.
4793 Havildar Nihala.

" C " Company.

Subadar Saraj Din.
  16 Coy. QrMr. Havildar Daswandi Khan.
 167 L./Havildar Fazal Dad.
 823 Sepoy Fazal Ilahi.
1366 Sepoy Fazal Hussain.
2519 L./Naik Sharhab Khan.
Cook Khuda Bux.

" D " Company.

  41 Havildar Mussadi.
1285 Sepoy Baldev Singh.
1076 Sepoy Sohab Singh.
1434 Sepoy Devi Singh.
1537 Sepoy Rupa.
 486 Naik Hans Ram.

" H.Q." Wing.

 112 Havildar Nand Singh.
 592 L./Naik Mohan.
1106 L./Naik Lohb Singh.
1737 L./Naik Walyat Khan.
2781 Sepoy Siri Ram.
2923 Sepoy Chotey Lall.
1745 Naik Dilbagh Singh.
1301 Havildar Niaz Ali.

APPENDIX X.

LIST OF MUSKETRY AND OTHER TROPHIES.

MUSKETRY.

DESCRIPTION.	MATCH.	DATE.
Plain tall cup, fluted stem	Inter-Regimental B.P.R.A., Native Army	1889
Do.	Do.	1890–91
Large double-handled round bowl	Do.	1892–93
Tall chased cup, two handles	Do.	1893
Oak shield in imitation of original shield, with miniature shields mounted on it	" Honour and Glory " Match, British and Indian Inter-Regimental	1894
Two claret jugs	Inter-Regimental, B.P.R.A., Native Army	1894
Boat-shaped cup	Do.	1894
Indian silver-handled jug, with camel on lid	C.-in-C.'s Match, B.P.R.A.	1897
Indian silver cup	Inter-Regimental, Native Army, B.P.R.A.	1899
Programme stand	A.G.'s Cup, Extended Order Competition, B.P.R.A., Meerut	1900
Font-like cup with handles (small)	Meerut Cup, B.P.R.A. ...	1900
Do. (larger)	Cawnpore Woollen Mills Cup (Regimental team)	1901
Cigarette lighter	Peshawar District Rifle Meeting	1902
Large low cup, two handles	Meerut Cup, B.P.R.A. ...	1905
Four silver pillar lamps ...	Attack practice, Meerut B.P.R.A.	1905
Large round bowl with rifles at sides	Cawnpore Woollen Mills Cup, won by No. IV D.C. Used as inter-officer challenge cup	1906–07
Three-handled bowl ...	Empire Day Challenge Cup	1910
Small two-handled bowl ...	Kohat Brigade Rifle Meeting, presented by Brig.-Gen. Birdwood	1911
Gold medal	Magdala Gold Medal, won by Hav. Allah Din	1911

MUSKETRY AND OTHER TROPHIES TOGETHER WITH PRESENTATION PLATE, OFFICERS' MESS, 2/15th PUNJAB REGIMENT, JHELUM, 1923.

DESCRIPTION.	MATCH.	DATE.
Small cup	The " De Kun " Cup—China—won by Mehr-bat	1912
Small cup	The " Gaupp Cup," S.C.R.A., won by Gul Badshah	1912
Silver statuette (Afridi in uniform)	Empire Day Cup, won by No. IV D.C.	1912–13
Cup	The Hatton Cup—China—Inter-Regimental in Hong-Kong, won by No. IV D.C.	1913–14

HOCKEY.

Cigarette lighter	Malakand F.F. Hockey Tournament. Presented by Capt. A. H. Macmahon	1900–01
Large plain cup	Peshawar Christmas Week Hockey Tournament	1910
Bowl on three dragons ...	Kowloon, U.S.R.S., Hockey Tournament	1913
Shield, silver on oak ...	Rawal Pindi District Hockey Tournament (Challenge)	1923–24

ATHLETIC CUPS WON IN 1924. (*All Challenge Cups.*)

Small-handled Cup ...	Jhelum Brigade Sports Championship.
Large round bowl with eight turrets	Rawal Pindi District : Relay race.
Tall and massive cup, surmounted by two colours	Rawal Pindi District : Uniacke Cup, Light-weight tug-of-war.
Large cup	Rawal Pindi District : Heavy-weight tug-of-war.
Large cup	Rawal Pindi District : Cross-country.